GENETIC RESOURCES OF INDIAN MAJOR CARPS

P V G K Reddy

Reprinted by arrangement with the
Food and Agriculture Organization of the United Nations

by
Daya Publishing House
A Division of
Astral International Pvt Ltd
New Delhi - 110 002

“This book was originally published by the Food and Agriculture Organization of the United Nations (FAO) as Vaccine Manual, FAO Animal Production and Health Series No. 35”.

“The designations employed and the presentation of material in this publication do not imply the expression of any opinion, whatsoever on the part of the Food and Agriculture Organization of the United Nations concerning the legal status of any country, territory, city or area or of its authorities, or concerning the delimitations of its frontiers or boundaries.”

‘The designations ‘developed’ and ‘developing’ economies are intended for statistical convenience and do not necessarily express a judgment about the stage reached by a particular country, territory or area in the development process.’

First Indian Edition, 2005
Reprinted, 2024
ISBN: 9788170353935

Published by : **Daya Publishing House**
A Division of
Astral International Pvt Ltd
4736/23, Ansari Road
Darya Ganj, New Delhi - 110002
Phone: 011-43549197
Email: info@astralint.com
Web: www.astralint.com

Printed at : **Chawla Offset Printers,**
Delhi-110 052

PREPARATION OF THIS DOCUMENT

This document was prepared within the framework of the Regular Programme activities of the Inland Water Resources and Aquaculture Service of the Fishery Resources Division. The primary objective of this document is to characterize the genetic resources of Indian major carps, especially in regard to aquaculture and fisheries. The document is primarily directed at resource managers, aquaculturists and scientists.

The original manuscript was prepared by Dr P.V.G.K. Reddy of the Central Institute of Freshwater Aquaculture, Kausalyaganga, Bhubaneswar, India. The manuscript was reviewed and edited by Dr D. Bartley.

The support of the Indian Council of Agriculture Research in allowing Dr Reddy to undertake this work is gratefully appreciated.

Reddy, P.V.G.K.
Genetic resources of Indian major carps.
FAO Fisheries Technical Paper. No. 387. Rome, FAO. 1999. 76p.

ABSTRACT

This publication, written for resource managers, aquaculturists and scientists, reviews the genetic resources of Indian major carps: the catla (*Catla catla),* the rohu (*Labeo rohita),* the mrigal (*Cirrhinus mrigala)*, and the kalbasu *(Labeo calbasu)*. The review includes information on spawning behaviour and breeding under natural and culture conditions, distribution, genetic characterization, status of genetic resources, conservation efforts, and hybridization among Indian major carps and with other cyprinids. Research on genetic improvement through selective breeding, chromosome-set manipulation and genetic engineering in India is reviewed. The status of Indian major carps in Bangladesh, Thailand, Vietnam and India work on cryopreservation and genebanking is also discussed.

Distribution :

Directors of Fisheries
FI Regional Offices
Fisheries Department
INGA members
Author

FAO Regional Offices
Fisheries Department
FIRI Members
Author

CONTENTS

1. INTRODUCTION

India being an agricultural country, its economy also mainly depends on the successful production of agricultural products. Besides, over 70 percent of Indian population inhabit rural areas. Until 1960s, India was not self-sufficient in producing food grains and had to depend on other countries for filling the gap between inadequacy and self-sufficiency. To overcome this situation and to feed its ever increasing population, the country had to struggle its way to self-sufficiency for producing food grains. To achieve this goal, agriculture scientists in the country were at the task from 1950s and by mid 1970s India could achieve self-sufficiency in producing food grains through developing genetically superior high yielding varieties of wheat and paddy and other vegetable products of the green-revolution. Simultaneously there were also efforts to improve and increase the farm animal products and an appreciable rise in the yield of milk and eggs production has been witnessed since the past one and a half to two decades.

Coming to the fisheries sector, India possesses nearly 11 percent of the world's 20,000 known species of fish, and is one of the richest nations in the world with regard to genetic resources of fishes which are distributed over a network of perennial river systems in the country. With regard to culture fisheries or aquaculture sector, though it appears that fish culture in India is almost as old as in China, the average per hectare production per year remained as low as 0.6 tones until the1960s. This was due to lack of proper technology and fish farmers that had to follow traditional or empirical methods for their fish farming.

The aquaculture systems in India and its neighbouring countries such as Bangladesh and Pakistan, mainly constitute Indian major carps viz. the Catla (*Catla catla*, Hamilton), the rohu (*Labeo rohita*, Hamilton), the mrigal (*Cirrhinus mrigala*, Hamilton) and some times the Kalbasu (*Labeo calbasu* , Hamilton). These carps contribute approximately 75 percent fo the aquaculture production in India (FAO, 1997). As these carps are economically very important, research on cultivating these species of fish was initiated in India during early 1950s to study and understand the biology of these economically important species, particularly the major carps and to develop suitable technologies for various farming systems.

FAO (1972) has already pointed out that application of genetic principles to the selection of breeding stocks in commercial aquaculture lags far behind plants and other animals breeding. The main reasons for this according to FAO appear to be 1) lack of complete control by man of the reproduction phases 2) paucity of facilities for carrying out selection programmes 3) shortage of aquaculturists with genetic training and 4) lack of awareness among aquaculturists of the importance of genetic work. FAO considers that considerable time may elapse before adequate solutions to the above problems are found particularly in many developing countries where there is an immediate need to improve fish production. Now after a lapse of over two decades considerable concern has been shown to use genetics in aquaculture and enhance aquaculture production in most of the developing countries which are discussed in this review.

Genetic work on Indian major carps was initiated in the late 1950s soon after the success of induced breeding of these carps through hypophyzation (Chaudhuri, 1959) with simple interspecific and intergeneric hybridization among these carps and later between these carps and Chinese carps. This was followed by cytogenetic work to study the karyotypes of major carps and their hybrids in a comparative sense and also to relate the viability and fertility of their hybrids. During early 1980s the modern genome manipulation (chromosomal engineering) work was initiated to develop techniques to induce and produce gynogens and polyploids to build up homozygous inbred lines and triploid sterile fish respectively. Biochemical genetic studies were also initiated more or less simultaneously. Towards late 1980s , with the experimental production of transgenic major carps (*Labeo rohita*), molecular genetics work was also initiated.

The first ever project on selective breeding of Indian major carps was initiated successfully in the early 1990s, in India at the Central Institute of Freshwater Aquaculture, Kausalyaganga (Bhubaneswar) in collaboration with the Institute of Aquaculture Research (AKVAFORSK), Norway.

Recently some laboratories in India and other developing countries in the region have initiated developing methods to identify genetic markers, particularly in major carps through allozyme, mitochondrial DNA (mt DNA), random amplified polymorphic DNA (RAPD), restriction fragment length polymorphism (RFLP) and microsatellite techniques, for the genetic characterization and population genetic studies.

2. GENETIC RESOURCES OF INDIAN MAJOR CARPS, THEIR DISTRIBUTION AND CHARACTERIZATION

2.1 Genetic resources of Indian major carps

The natural genetic resources of Indian major carps come from the network of the Ganga river system, the Sindh and the Brahmaputra river systems in the north and the east-coast and west coast river systems flowing through in the south and central India respectively. The major river system of India have been given in Fig. 1.

2.1.1 About Indian major carps

The major carps of India fall under three genera , *Catla, Labeo* and *Cirrhinus*. Under the genus *Catla*, the species *C. catla*, under the genus *Labeo* fall the species *L. rohita, L calbasu, L. fimbriatus, L. bata, L. gonius,* and under the genus *Cirrhinus* fall the species *Cirrhinus mrigala, C. reba, C. cirrhosa*. Interspecific hybridization work has been carried out between the species of the genus *Labeo* and intergeneric hybridization between the species of the above mention three genera.

Due to their fast growing nature and taste, Indian major carps enjoy a prime position in the Indian aquaculture scenario. These highly prized fishes though originally inhabitants of the Ganga river network in North India and the rivers of Pakistan, Bangladesh, Nepal and Burma, are also transplanted into other rivers in central as well as peninsular India. Besides this, these carps have been imported by many other countries such as Thailand, Vietnam, Laos (Welcomme, 1988).

Catla feeds mainly on Zooplankton at the upper zone of the water body and rohu mostly feeds on periphytonic (phytoplankton) forms found attached to submerged vegetation and other objects occupying the water column , while mrigal and kalbasu feed at the bottom on bottom biota such as tubifex and other blood worms. Thus these carps have mutually compatible and complimentary food habits. Major carps are shown in Plate 1.

Taking advantage of the mutually compatible and complimentary food habits of these carps, the present day intensive and extensive polyspecies or multispecies culture has been developed from 1970s, which is popularly known as the composite carp culture. This technology has developed and has demonstrated a gradual increase in the fish production from the culture systems from 3-4 t/ha/year to 10-15 t/ha/year. (Chaudhuri *et al.*, 1974; 1975; 1978 and Tripathi *et al.*, 1997).

These technologies are mainly based on manipulation of species ratios to exploit the available natural food in all the niches of the culture medium and supplementary feeding and manuring. Other management measures such as water replenishment, health monitoring etc. are also part of the package.

Carp culture practice which was hitherto confined mostly to the eastern region of the country has become popular almost all over the country with the development of modern aquaculture technologies. Consequently the demand for seed of major carps gradually started increasing. With the increase in the demand, the hypophyzation technique has spread all over the subcontinent and became popular among many fish farmers. As a result the production of seed has increased.

With the epoch making success in inducing these carps to breed through hypophyzation during late 1950s (Chaudhuri and Alikunih, 1957), the dependence on riverine sources for the seed of these carps has been gradually reduced. Now, hypophyzation, has been adopted by many of the rural fish farmers to produce the seed in adequate quantities.

However, in the absence of proper breeding plans, this has led to a gradual decline in the genetic quality of the seed. Consequently the negative effect of inbreeding started appearing gradually with the characteristic poor survival and slow growth, besides disease susceptibility of the hatchery produced seed (Ibrahim *et al.*, 1982). All this suggests that it is an appropriate time to act seriously about the genetic improvement of these carps.

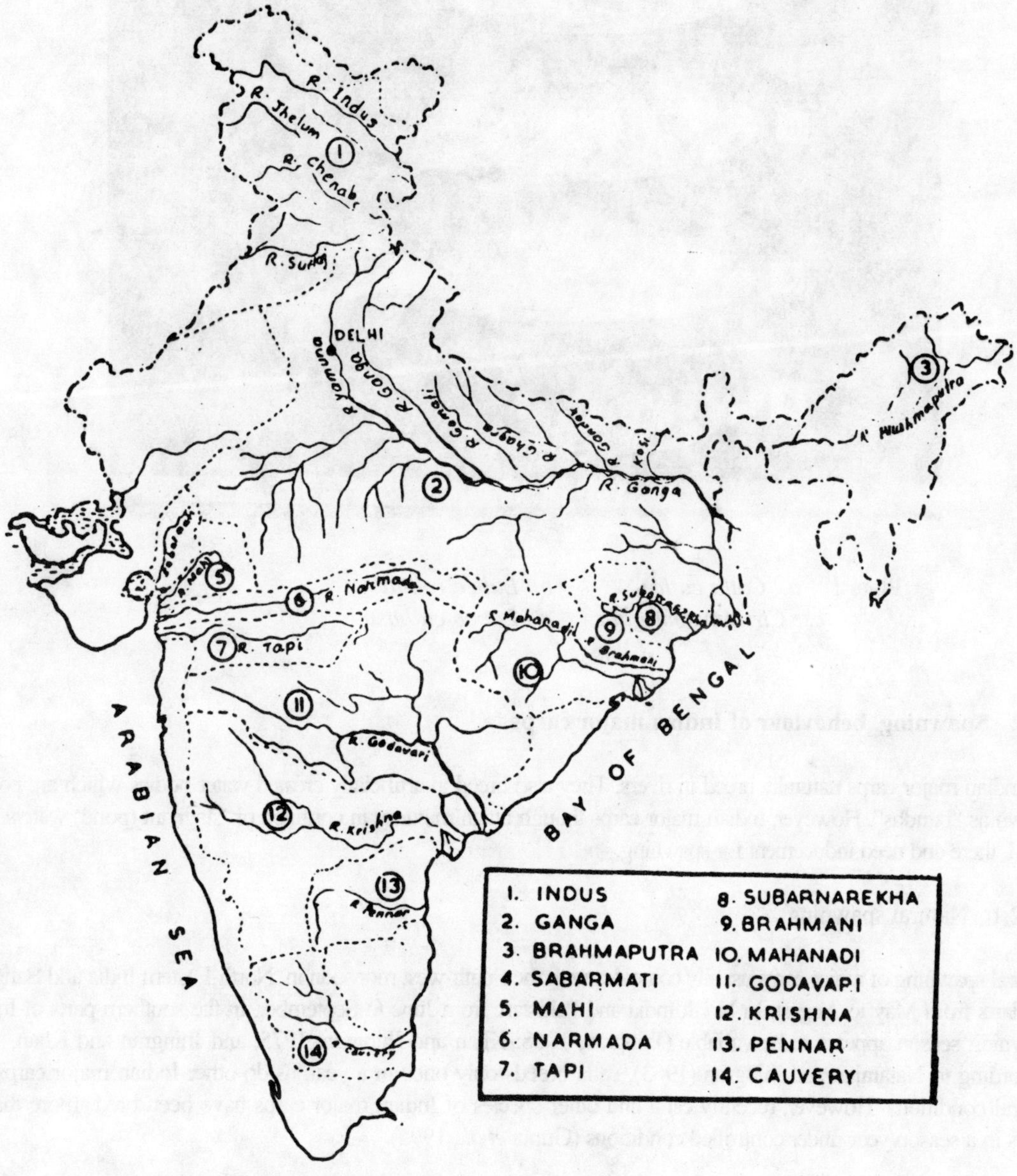

Figure 1. Major rivers of India (Jhingran, 1984)

Plate 1. *a. Catla catla* *b. Labeo rohita*
c. Cirrhinus mrigala *d. Labeo calbasu*

2.1.2 Spawning behaviour of Indian major carps

All Indian major carps naturally breed in rivers. They also breed in artificially created water bodies which are popularly known as "bundhs". However, Indian major carps though attain maturity in confined or stagnant (pond) waters, do not breed there and need inducement for spawning.

2.1.2.1 Natural spawning

Natural spawning of major carps usually coincides with the South-west monsoon in North-Eastern India and Bangladesh and lasts from May to August. In North India and Pakistan, from June to September. In the southern parts of India the spawning season appears to be variable (Jhingran, 1968; Khan and Jhingran, 1975; and Jhingran and Khan, 1979) . According to Natarajan and Jhingran (1963), catla breeds only once in a year, as do other Indian major carps under natural conditions. However, recently catla and other species of Indian major carps have been bred more than four times in a season/year under controlled conditions (Gupta *et al.*, 1995).

Regarding the requirements for spawning and the nature of spawning grounds, the observations of various workers differ slightly. Some are of the view that shallow areas which are inundated by heavy monsoon floods form the breeding grounds of these fish (Hora 1945, Hussain 1945 and Khanna 1958). Others viewed that it is only the availability of shallow spawning grounds that is the important factor for spawning (Khan, 1947; Ganapati and Alikunhi, 1950).

Again regarding the nature of the spawning ground, observations differ. In rivers Godavari, Krishna and Cauvery in South India, sections having large submerged rocks or emergent vegetation form the spawning grounds for catla (Chacko, 1946; Chacko and Kuriyan, 1948a; and Menon *et al.*, 1959). Job and Chacko (1948) described as many as 22 spawning grounds. Spawning of major carps was reported over hard and sandy beds and also on rocky submerged embankments of rivers, reservoirs and "bundhs" in Madhya Pradesh (Dubey and Tuli, 1961).

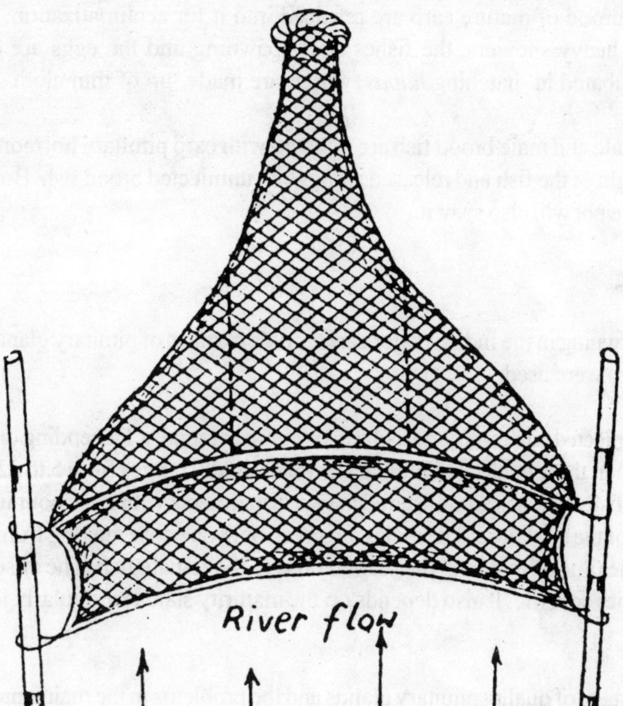

Figure 2. Spawn collection net (from K.K. Ghosh, 1972)
Net for collecting spawn from the rivers.
(Note: the broaden end (mouth) of the net facing towards the flow of the river indicated with arrows)

Riverine fish seed is collected with the help of specially designed, funnel shaped nets (Jhingran, 1968). It consists of two parts, i.e. the net proper and a detachable tail piece locally called as "gamcha" which is fixed to the cod end. The rear of the net tapers to the narrow end. The net is fixed with the broader (mouth) end facing the water current. The spawn/fry are collected at the tail piece. Usually a number of such nets are fixed, depending on the favorable conditions. Figure 2.

2.1.2.2 "Bundh" breeding

Bundhs are special type of impoundments which are mostly found in the states of West Bengal, Madhya Pradesh, Rajasthan and Bihar. These are of two types, Perennial (wet bundh) and seasonal (dry bundhs). A typical wet-bund is usually located in the slope of a vast catchment area of undulating terrain, having proper embankments and with an inlet towards the upland and an outlet on the opposite side. It has a deeper and vast shallow area which dries up in summer. The deeper portion of the bundh retains water during summer where an adequate stock of carps is maintained. During monsoon when there is a heavy shower, the rain water from the catchment area rushes through the streamlets and submerges a major portion of the bundh while the excess water flows out through the outlet.

On the other hand a *"dry-bundh"* is like a seasonal pond and much smaller and shallower than the *wet-bundh*. For most part of the season the *dry-bundh* remains dry. When sufficient water accumulates with the first showers of monsoon rains brood of mature carp are released into it for acclimatization. Later when the bundh gets flooded with the next heavy showers, the fishes start spawning and the eggs are collected by means of mosquito net cloth and incubated in hatching *hapas*, which are made up of thin cloth .

Sometimes a few of the female and male brood fish are injected with carp pituitary hormones at a nominal dose of about 5-10 mg/kg body weight of the fish and released along with uninjected brood fish. Both fish that received the hormones and those that have not will also spawn.

2.1.2.3 Controlled breeding

In early efforts to induce spawning in the Indian major carps, crude extracts of pituitary gland preferably of the same species (homoplastic glands) were used.

The pituitary hormones are injected either intramuscularly or interperitonially. Depending on the condition of brood fish and the prevailing weather, the total dose of pituitary extract usually ranges from 5 to 22 mg per kilogram body weight in the case of female fish, which is administered in two installments with a six-hour interval. The quantity of first dose is nearly one third of the total dose. The males are given only one dose ranging from 2-3 mg of pituitary per kilogram body weight, at the time of injecting the second dose to the female fish. The dose of pituitary hormones however, appears to be species specific. It also depends on the maturity status particularly in the case of the female brood fish.

In the recent past, due to shortage of quality pituitary glands and the problems in the maintenance of donor fish, many substitutes have come into use to replace pituitary hormones, such as Human Chorionic Gonadotropin (HCG), Salmon gonadotropin (partially purified), Luteinizing Hormone-Releasing Hormone (LHRH) and its Analogues, Gonadotropin Releasing Hormone (Gn RH) etc. However, presently a formula of S-GnRH and domperidone which is marketed in the name of OVAPRIM, is much in use to induce the major carps to breed. OVAPRIM is used mostly for female brood fish @ 0.3-0.4 ml/kg body weight of the fish in a single dose.

With pituitary extracts the female fish will be ovulating within 5-6 hours after receiving the second dose, whereas OVAPRIM takes 6-7 hours to induce ovulation. The release of milt by the males usually synchronized with the ovulation of females. For genetic experiments, stripping procedure is generally followed.

2.2 Distribution

The Indian major carps which are also known as the gangetic carps are the natural inhabitants of the Ganga river network, namely the Ganga, the Gomati, the Yamuna and the Brahmaputra and the Indus river system in North India.

The Ganga river system running to a total length of about 8, 047 km, besides the major carps, harbours the richest freshwater fauna of India, ranging from mahaseers and the torrential fishes of hills to a wide array of other fishes of great commercial value.

The Brahmaputra river system having a length of 4023 km consists of rich fish fauna of torrential streams in its upper reaches, however the fish are not of much economic value. Its middle and lower reaches however, have several species of carps, catfishes and other commercially important species of fish.

Regarding the Indus system, only part of a small segment is in India, the rest being in Pakistan. The important rivers of this system are the Beas and the Sutlej. These rivers harbour not only the indigenous carps in the lower sections but also the exotic rainbow and brown trouts in their upper reaches.

The main Rivers of the East coast river system in the peninsular India are the Mahanadi, the Godavari, the Krishna and the Cauvery. The Mahanadi naturally has all the species of Indian major carps, which are common in the Ganga system. Rivers Godavari, Krishna and Cauvery have their own several indigenous carp species, which are referred as minor carps such as *Labeo bata, L. gonius, L. fimbriatus, Cirrhinus reba* etc. However, the major carps have been repeatedly transplanted into these rivers and have become established in them, contributing significantly to the fish fauna of these rivers.

The major rivers of the West coast river system are the Narmada and Tapti. Their fish fauna consists of carps, catfishes, mahaseers, murrels, purches, prawns etc.

Distribution of major carp resources also includes that part of the Indus river system which falls into Pakistan after its formation and tributaries of Brahmaputra flowing through Bangladesh.

2.2.1 Distribution of *Catla catla* (catla)

Jhingran (1968) described the distribution of *Catla catla*, which starts from the Ganga river network in the north to the Krishna river down south of India, Pakistan, Bangladesh and Burma. It is also found in Nepal (De witt, 1960 as referred by Jhingran, 1968).

The fingerlings of catla, introduced into the Cuddapah-Kurnool canal from river Godavari in 1909 found their way into river Penna and the connected waters in the Nellore district of the state of Andhra Pradesh. Catla fingerlings were also introduced in Cavery river during 1920s and subsequently into the Bhavani (Hornel, 1924). Later the species was introduced into Periyar lake, Powai lake (Chacko, 1948 and Goldschmilt, 1953). The species has become widely established (Fig.3).

Jhingran (1968) has also mentioned that catla fingerlings have been also exported to Israel in 1954 and to Japan, and Mauritius in the 1960s. According to FAO Database on Introduction Aquatic Species (DIAS) (www.FAO.Org.) catla has been distributed to other countries also such as Zimbabwe, Israel, Bhutan, Philippines, Former USSR, Japan, Sri Lanka, Laos, Pakistan, Malaysia, Thailand, Vietnam and Mauritius.

2.2.2 Distribution of *Labeo rohita* (rohu)

The main source of literature about the distribution of rohu is Day (1878 and 1889). He reported that the species is distributed in freshwaters of Sindh and Punjab (Pakistan), through India, Assam, Bangladesh and Burma. He also stated that rohu is not found in Madras (South India).

Later studies of other workers (Setna and Kulkarni, 1949) and Alikunhi and Chaudhuri, 1951 and Mishra, 1959) have mentioned the occurrence of this species in many other places, viz., Sabarmati drainage, in the rivers Narmada, Tapti, Godavari, Mahanadi etc. though it is more common in plains of North India.

Its distribution has been also mentioned in other neighbouring countries such as Bangladesh, Burma and Nepal. Khan and Jhingran (1975) have given other details on the distribution of rohu through transplantation. However, according to FAO DIAS (19) besides the above mentioned countries the distribution of rohu has also extended to the Philippines, Former USSR, Japan, Sri Lanka, Laos, Pakistan, Malaysia, Thailand, Vietnam, Madagascar, China and Mauritius (Fig. 4).

2.2.3 Distribution of *Cirrhinus mrigala* (mrigal)

The earliest work on distribution of mrigal appears to be Day (1878 & 1889). According to him mrigal inhabits rivers and tanks (water bodies much larger than ponds) in Bengal (undivided which includes Bangladesh) Deccan, North-West provinces, Punjab, Sindh (Pakistan), Cutch (partly in Pakistan and partly in Gujarat, Rajasthan provinces of India) and Burma. Later workers (Alikunhi, 1948 and 1957, and Mishra, 1959) recorded its distribution in the major river systems of India, including river Godavari in the south.

In addition, Mishra (1959) also mentioned about its distribution in Pakistan, Bangladesh and Burma. The species was also recorded from Nepal (De Witt, 1960). According to FAO DIAS (19) besides the above countries, mrigal has

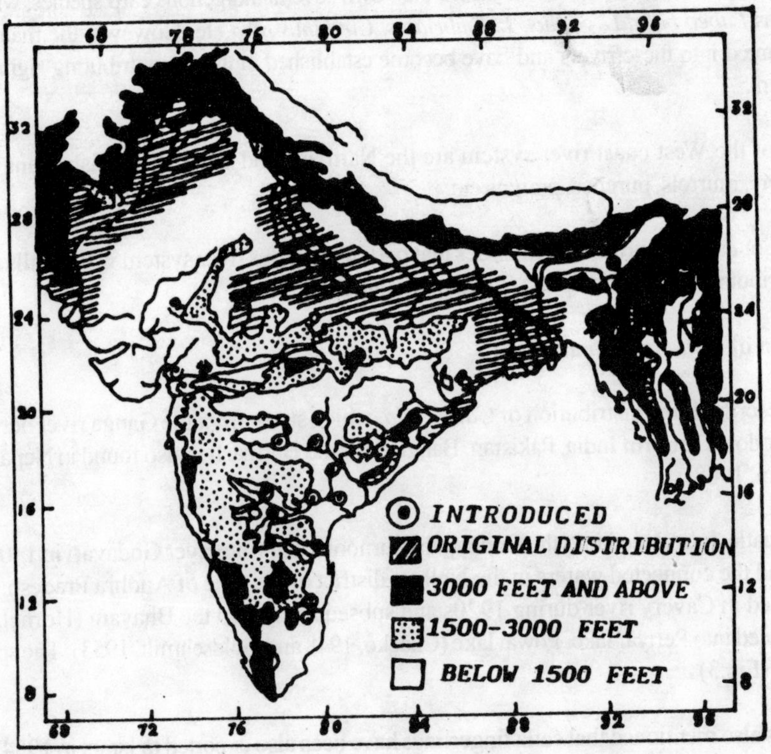

Figure 3. Geographical distribution of *Catla catla*. Crosses indicate areas where the fish has been successfully transplanted. From : Jhingran, 1968 (with FAO's permission)

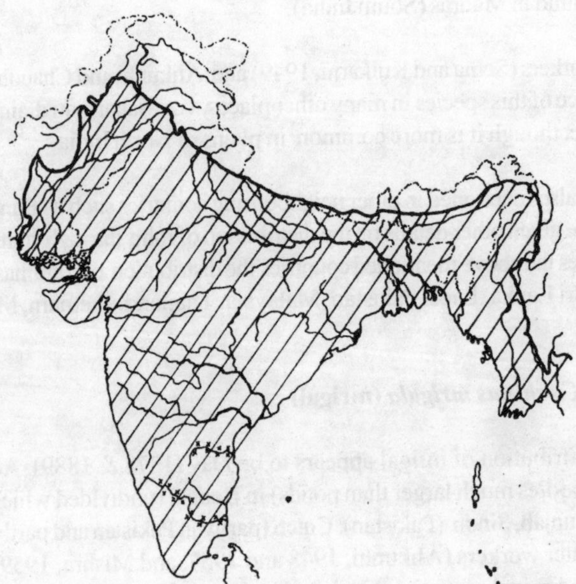

Figure 4. Geographical distribution of *Labeo rohita*. From Khan and Jhingran, 1975 (with FAO's permission)

also introduced to Bhutan, Philippines, Former USSR, Japan, Sri Lanka, Laos, Malaysia, Thailand, Vietnam and Mauritius (Figure 5).

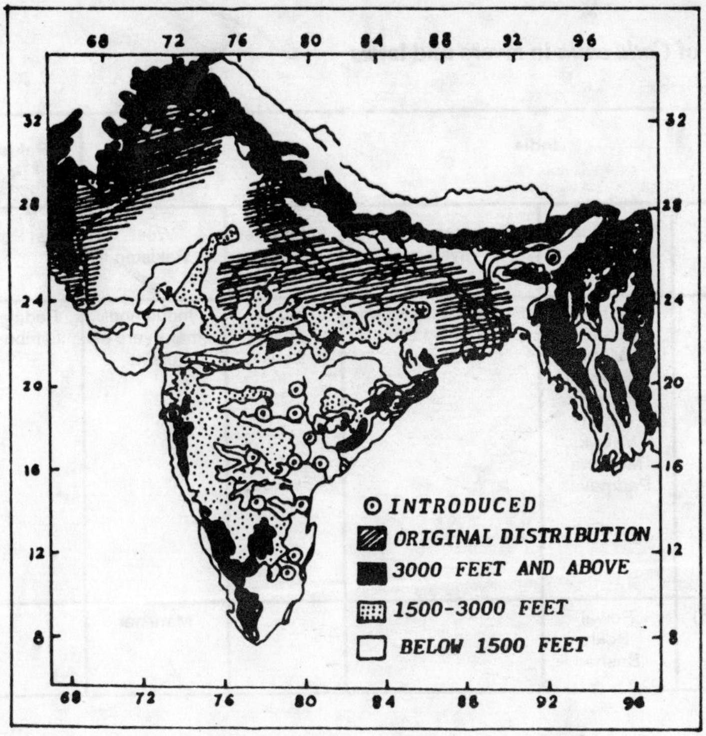

Figure 5. Geographical distribution of *Cirrhinus mrigala*. From Jhingran and Khan, 1979 (with FAO's permission)

2.2.4 Distribution of *Labeo calbasu*

Like the other three major carp species. *L. calbasu* also enjoys wider distribution in almost all those countries namely, Pakistan, India, Bangladesh, Burma, Thailand and also in Yamuna (South China) . Day, 1878 mentioned that *L. calbasu* is distributed in Punjab, Sindh, Cutch Deccan, South India and Malabar, from Krishna through Orissa, Bengal and Burma. In India, according to Jhingran (1984) the species is found in streams of Punjab, West Bengal, Orissa and parts of South India. It is more abundant in rivers, above tidal reaches and lakes.

It is an important food fish and at several places it is referred as "lake rohu". It is an important game fish too in tanks . It has been reported that *L. calbasu* thrives better in tanks and lakes than in running waters (Talwar and Jhingran, 1991).

Distribution pattern of catla, rohu and mrigal in rivers of different countries has been shown in Tables 1, 2 and 3 respectively.

2.3 Major carps in reservoirs

Reservoir here refers to the impoundments formed as a result of erecting dam on a particular river. There are a number of such reservoirs in India.

Srivastava *et al*. (1985), as referred by Sugunan (1995) have compiled a list of 975 large and medium reservoirs in India, with an estimated area of 1.7 million ha. These reservoirs have major carps as part of their ichthyofauna. All the species of major carps do not seem to thrive equally well in all the reservoirs. Again, there may be fluctuations in the abundance of the same species in the same reservoir.

In Stanley reservoir, in the state of Tamil Nadu, catla was dominating in the fish landings during 1960-'65, but it was reduced to 10% during 1969-'70 to 1972-73. Mrigal, introduced into the Stanley reservoir during 1950-'51, first appeared in the catches during 1957-'58 and contributed over 13% of the catch in 1966-67. However, later catch declined to insignificance. Rohu also showed some initial increase but later almost followed mrigal (Sugunan, 1995). The trend in most of the reservoirs more or less appears the same. The reasons for the dominance of a particular species at one time and the decline of the same species later seem to be due to recruitment failure from the water level fluctuations and predator pressure (Sreenivasan, 1966). Moreover, other aspects like the availability and supply of the required natural food of a particular species may play a role in the establishment and thriving of the species in reservoirs.

Table 1. Distribution of *Catla catla* in rivers and lakes

Habitat	Buma	India						Pakistan
		Ganga river system	East Coast system	West Coast system	Brahmaputra river system	Indus river system	West Pakistan	East Pakistan
Rivers and important tributaries	Pegu	Ganga Yamuna Ghaghra Gomti Rapti Sarda Ramganga Kosi Sone Damodar Chambal Betwa Ken	Mahanadi Godavari Manjra Krisna Tungabhadra Suvamarekha Uraongarha southkoel Cauvery	Narmada Tapti Mahi Mindhola hathmati Khari Vatrak Meshwa Padmavati	Brahmaputra Kalang Burhi Dhing Dhansiri Dhiko	Indus Sutlej Beas	Indus and other rivers of plains	Padma and its tributaries
Lakes	Indawgyi	Ranchi Bhopal Lower	Sur (near Puri) Pennar	Powai Bokh Bashan			Manchar	

Only such rivers have been mentioned from where mrigal has been especially reported in ichthyological literature. From : Khan and Jhingran, 1979 (with FAO'S permission).

Table 2. Distribution of *Labeo rohita* in rivers and lakes

Habitat	Burma	India					Bangladesh	Pakistan
		Ganga river system	East Coast system	West Coast system	Brahmaputa river system	Indus river system		
Rivers and important tributaries	Irrawady Myintze Panhlaing Sittang	Ganga Yamuna Ghaghra Gomati Rapti Sarda Ramganga Kosi Son Damodar Chambal Betwa Ken	Mahanadi Godavari Krisna Cauvery	Narmada Tapti Mahi Sisodra Sabarmati	Brahmaputra Kalang Burhi Dhing Dhansiri Dhiko	Indus Sutlej Beas	Padma and its tributaries	Indus and other rivers of plains
Lakes	Indawgyi	Ranchi Lower Bhopal		Powai Bokh				

Only such rivers have been mentioned from where rohu has been especially reported ichthyological literature.

From: Khan and Jhingran, 1975 (with FAO's permission).

Table 3. Distribution of *cirrhinus mrigala* in rivers and lakes

Habitat	Burma	India					Bangladesh	Pakistan
		Ganga river system	East Coast system	West Coast system	Brahmaputra river system	Indus river system		
Rivers and important tributaries	Irrawady Myintze Panhlaing Sittang	Ganga Yamuna Ghaghra Gomati Rapti Sarda Ramganga Kosi Son Damodar Chambal Betwa Ken	Mahanadi Godavari Krisna Cauvery	Narmada Tapti Mahi Sisodra Sabarmati	Brahmaputra Kalang Burhi Dhing Dhansiri Dhiko	Indus Sutlej Beas	Padma and its tributaries	Indus and other rivers of plains
Lakes	Indawgyi	Ranchi Kolleru Lower Bhopal	Kolleru	Powai Bokh				

Only such rivers have been mentioned from where mrigal has been especially reported in ichthyological literature.
From: Khan and Jhingran, 1979 (with FAO's permission).

However, with careful management and monitoring, reservoirs in India may form a potential resource for not only major carps but also many other commercially important species.

2.4 Genetic characterization of major carps

Application of genetics in fish farming is relatively of recent origin. Until recently no attempts have been made to genetically characterize any of the major carp species. No reports are available on whether there exist genetically different populations among the members of a given species of major carps in any of the major river systems mentioned earlier. The main reasons for this seem to be prioritization of problem/tasks in the fishery sector. As mentioned earlier, the priorities in fisheries, not only in India but also in other developing countries in Asia were different, ever since systematic research has been initiated. The need was to identify the economically important species of fish and study their biology and methods of propagation under controlled conditions. Once this was established, the fishery workers were busy in developing, technologies for carp/fish farming systems to boost the production. Except for the work on hybridization, no importance was given to any aspects of genetics what so ever, during these years.

The composite or multispecies culture technologies so far developed are based on species manipulation and application of certain management practices. These technologies, no doubt have boosted the fishculture in India several folds. However, at present it is felt that any further improvement in fish production may not be possible with these technologies and the researcher gradually started realizing the importance of other aspects such as genetic quality and improvement of the candidate species by fully exploiting their hitherto untapped genetic potentials. The urgency of acquiring knowledge about the existence of different fish stocks/populations among the different species of the economically most important fish (carps) is now felt. Lack of appropriate and accurate methodologies for genetic characterization and identification at population/species level has also been one of the main constraints to full use of genetic resources. The various methods available earlier, much before the advent of biochemical and molecular (DNA) techniques for stock identification or to study the existence of different populations in a given species were only the morphometric measurements and meristic counts. These methods however, do not provide the degree of polymorphism distinguishable by modern methods especially within species.

According to Jhingran (1968) no distinct races or varieties of catla are known. This author has also given an account of the morphological traits of catla as described by different authors. However, Sinha and Khan (1989) mentioned that in Rihand Dam (Madhya Pradesh), India, the occurrence of three intra-specific populations within catla, each of which can be distinguishable morphologically by short, medium and long pectorals and also by specific ecological roles.

Khan and Jhingran (1975) mentioned that Khan (1972) conducted some studies on rohu collected from two different environments, moats, representing lentic environments and from the rivers Ganga and Yamuna representing lotic environments. The analysis of the data has shown significant differences in each of the characters suggesting that the fishes of moats and rivers belonged to independent stocks. Except for these sporadic reports which are mainly based on morphological and other phenotypic characters, no other reports are available regarding the existence of different populations or races among the species of Indian major carps.

Later developments include the cytogenetic, biochemical and molecular genetic methods. Among these, the biochemical genetic techniques, especially the use of isozyme electrophoretic markers, were in extensive use in fishes, through the 1960s and 1970s. However, since isozyme pattern cannot detect point mutations and conservative amino acid substitutions, they do not reveal all of the genetic variation actually present (Padhi and Mandal, 1995).

At the National Bureau of Fish Genetic Resources (NBFGR) at Lucknow (India) a number of species including major carps have been cytogenetically investigated (Ponniah, 1997). Indian major carps have the same diploid number of 50 chromosomes and no intraspecific variation in chromosomal number has been detected. Similarly, no intraspecific Nuclear Organizing Region NOR variation have been observed in both riverine and hatchery populations of *L. rohita*.

A relatively recent advancement made in the area of molecular genetics through developing molecular DNA markers has provided more accuracy for the identification of fish stocks in the form of restriction fragment length polymorphism (RFLP) of mitochondrial DNA (mt DNA) and nuclear DNA. Genetic polymorphism in nuclear DNA can also be detected by using random olegonucleotide primer for polymerase chain reaction PCR amplification followed by gel electrophoresis. This is known as random amplified polymorphic DNA (RAPD), which can be applied to genetic stock identification of quantitative traits loci (QTL) (Padhi and Mandal, 1995a). These techniques are very sophisticated and involve high cost infrastructure, which is often not within the reach of many developing countries. This appears to be one of the main reasons why the studies on stock identification in fishes of many developing areas have not been carried out on full scale until recently. However, use of the DNA technology is gaining momentum since 1990s. Many laboratories and institutions have come up in fisheries and are involved in research in this direction.

2.4.1 Application of biochemical and molecular genetic techniques

Information on some recent investigations with regard to biochemical and molecular genetic aspects on Indian major carps is discussed below.

2.4.1.1 Biochemical genetic studies

Biochemical genetic studies on Indian major carps through electrophoretic analysis of isozymes was initiated around the mid to late 1980s. These were mainly on the enzyme lactate dehydrogenase (LDH) patterns in developing embryos and in eye lens and red cell haemolysates (Padhi and Khuda-Bukhsh, 1989a,b), Ldh-C gene expression in liver tissue (Rao *et al.*, 1989), malate dehydrogenose MDH pattern in red cell haemolysate and similar electrophoretic patterns in muscle proteins (Padhi and Khud-Bukhsh, 1989b). These studies have shown that the three Indian major carps species namely *Catla catla, Labeo rohita* and *Cirrhinus mrigala*, in many of their biochemical characters were found to be very similar.

Earlier to this, Sahoo (1987) worked on the plasma, haemoglobin and transferrin pattern of two Indian carp hybrids; one produced by crossing male *L. rohita* with female *C. catla* and the other between male *L. rohita* with female *Cirrhinus mrigala*. All the plasma protein bands of both the parents *L. rohita* and *Catla catla* were expressed in their hybrid.

Regarding the haemoglobin pattern the hybrid between *L. rohita* and *C. catla* exhibited a broad intensely stained band which was similar to its parents. Similarly the hybrid between *L. rohita* and *Cirrhinus mrigala* also had a single band closely similar to parents.

With regard to transferrin bands, *L. rohita* has two and *C. catla* has one but their hybrid has inherited the extra band also from *L. rohita*. The hybrid of male rohu and female mrigal has shown two transferrin bands as in both the parents. Thus the pattern of plasma, haemoglobin and transferrin bands in the parent species were inherited by their respective hybrids.

Rao *et al*., (1989) have studied lactate dehydrogenase isozymes and taxonomic significance of Ldh C-gene in fifty two species of Teleostean fishes including Indian major carps. Liver specific Ldh C-gene expression was observed in *C. catla, Cirrhinus mrigala, L. rohita* and *L. calbasu*, also in many other species including Chinese carps, belonging to family Cyprinidae. Thus, members of the order cypiniformes show a dominance of liver-specific C-gene expression.

Padhi and Khuda-Bukhsh (1989a) studied the lactate dehydrogenase isozyme patterns in developing stages of four species of Indian carps. They found that usually the LDH gene expression during development tends to be similar in closely related species while the pattern may be dissimilar in more distantly related forms. It thus helps to understand the degree of evolutionary kinship among related organisms. The LDH banding latterns in the three species of major carps viz. *C. catla, Cirrhinus mrigala* and *L. rohita* was observed to be very similar and ali of them exhibited persistent pattern of three bands in all the developmental stages studied This indicates their genetic closeness (Padhi and Khuda-Bukhsh, 1989a).

Padhi and Khud-Bukhsh (1989b) have also studied malate dehydrogenase isozyme patterns during the development of five species of carps namely *C. catla, L. rohita, L. bata, Cirrhinus mrigala* and *Ctenopharyngodon idella*. Though significant changes in MDH pattern were marked during hatching, in the case of *C. catla* in the form of additional bands, no such MDH activity was apparent in the other species. The MDH patterns however, tended to fluctuate at different pre and post hatching stages in these carps, possibly reflecting differential gene expression (Padhi and Khuda-Bukhsh, 1989b). The number of MDH bands expressed during development, ranged between 1-6 in *C. catla*, 3-4 in *L. rohita*, 2-5 in *Cirrhinus mrigala*. While the number of MDH bands in *L. rohita* did not fluctuate much during pre-hatching stages, differential expression of MDH bands was noticed in other species.

To explain the genetic basis of expression, specific antisera and other confirmatory studies are required. Xanthine dehydrogenase isozymes (XDH) and its distribution in different tissues namely skeletal muscle, heart, liver, eye, kidney and brain and its taxonomic significance have been studied in *C. catla, Cirrhinus mrigala, L. rohita* and *L. calbasu* (Padhi and Khuda-Bukhsh, 1990a). In this study single band XDH phenotype was expressed uniformly in all the six tissues studied in *C. catla*. In *Cirrhinus mrigala* and *L. rohita* it was expressed in most of the tissues. However, in the case of *C. calbasu* and also *L. rohita* multiple XDH phenotypes showing four or five bands were also encountered in certain tissues.

In another study, Padhi and Khuda-Bukhsh (1990a) have found through simple gel electrophoretic study of the surface mucus of catla, rohu and mrigal, conspicuous differences which can differentiate the three species. The authors are of the opinion that a critical analysis of components of surface mucus may also be usefully exploited in population/stock identification of carps. The surface mucus proteins were expressed as seven bands in *C. catal*, three bands in *L. rohita* and five electrophoretically analysable components in *Cirrhinus mrigala* (Padhi and Khuda-Bukhsh, 1990a).

Basu *et al*., (1992) have studied the LDH isozymes in embryonic and post embryonic stages of *L. rohita, Cirrhinus mrigala* and *C. catla*, using starch gel electrophoresis. These studies have shown species specific and differential enzyme locus (gene) expression pattern in LDH up to 18th hour of study. The isozymes seemed to be completely active, 36th hour after fertilization and showed electrophoretic patterns very similar to those of the adults. It was also mentioned that comparative analysis of isozymes of the three species has also enabled species identification even during the embryonic stages which other wise would be impossible to identify on the basis of morphological characters only (Basu *et al.,* 1992).

Other studies in the area of biochemical genetics pertaining to Indian major carps are gel electrophoretic patterns of transferrin in the four Indian major carp species, *C. catla, Cirrhinus mrigala, L. rohita* and *L. calbasu* along with 30 other freshwater species of fish. While only one transferrin band was observed in *L. calbasu*, in *C. catla, Cirrhinus mrigala* and *L. rohita* expression of both one band and two band phenotypes were noticed (Barat and Khud-Bukhsh, 1992). Sahoo (1993) studied the plasma, haemoglobin and transferrin pattern in two intergeneric carp hybrids. The hybrids between male *L. rohita* x female *C. catla* and male *L. rohita* x female *Cirrhinus mrigala* have shown a haemoglobin pattern more or less similar to their respective parent species. However, the plasma band profile of the hybrids has shown a mixture of bands of the parent species. Again, the pattern of transferrin bands in the hybrids though resemble the parent species in the number and staining intensity, differed in mobility (Sahoo, 1993).

Isozymes studies have also been made to distinguish diploids and tetraploids. Sarangi and Mandal (1996) have made some isozyme studies in one of the Indian major carps namely *L. rohita* to distinguish diploid individuals from tetraploids. In general, intensification of isozyme bands has been observed in tetraploids. The polymorphism in (G6 PD-1 and 2) loci in eye lens, G6 PD-2 in kidney and G6 PD-1 in skeletal muscles and an esterase (E ST-1) locus in kidney could be used as reliable marker in identifying tetraploids from diploids (Sarangi and Mandal, 1996).

Chaudhuri and Gopala Krishna (1998) studied tissue specificity and degree of polymorphism of five enzymes systems in *L. rohita* from river Yamuna namely Aspartate transaminase AAT (E.C.2.6.2.2), Malate dehydrogenase MDH (E.C.1.1.1.3.7), Lactate dehydrogenase LDH (E.C.1.1.1.27), Malic enzyme (ME) and Glucose Phosphate Isomerase GPI (E.C.5.3.1.9) enzyme in liver, muscles, heart and brain tissues. It was reported that *MDH, ME* and *GPI* had two loci each in all tissues while *LDH* had trace loci with one specific to the liver. *AAT* with two alleles was the only polymorphic locus among the five enzymes studied.

Gel electrophoretic studies have been also carried out in over thirty five species of Indian teleostean fishes, which include major carps, in relation to plasma protein profiles (Khuda-Bukhsh and Sahoo, 1998). The plasma protein profile of *C. catla* revealed 9 bands, including a cluster of 4 linear bands near the origin. The number of bands in the cluster sometimes varied from 4-6. In *Cirrhinus mrigala* the electropherogram of plasma protein revealed 10 bands, including the cluster of bands near origin which generally contained 5 linear bands but sometimes 6-7 bands were also appeared in the cluster. In the case of *Labeo calbasu* and *L. rohita* 10 bands were observed in each case. (Khuda-Bhukhsh and Sahoo, 1998).

2.4.1.2 Molecular genetics studies

Methods for the rapid isolation of mitochondrial DNA (mt DNA) from fish have been developed (Padhi and Mandal, 1993), followed by mt DNA RFLP investigation studies for genetic stock identification (Padhi and Mandal, 1995), DNA fingerprinting (Majumdar *et al.,* 1997) and characterization of Mbol satellites (Padhi *et al.,* 1998) have been also carried out in major carps. It has been reported that mt DNA technique enables effective yield of good quantity of mt DNA without the use of ultracentrifugation. The authors claim that this method would be useful for preparing large number of mt DNA samples for studies pertaining to genetic stock differentiation, evolutionary analysis of related taxa and identification of maternity of the naturally occurring fish hybrids. The fish species *C. catla, L. rohita, L. bata* and hybrids of *C. catla* X *L. rohita* were used for this study.

The feasibility of actual application of mt DNA RFLP analysis methods have been tested by Padhi and Mandal (1995b) in Indian major carps for genetic stock identification by using *C. catla, L. rohita* and *L. calbasu* from both riverine (river Ganga) and farm sources. Restriction endonucleage analysis of mitochondrial DNA in *L. rohita* from the river Ganga and hatchery maintained (farm) population revealed polymorphism at the Hind III restriction site. The mt DNA genome size in rohu has been reported to be about 17.8 kb. Padhi and Mandal (1995b) also reported that *C. catla* and *L. calbasu*, collected from both farm and riverine sources, when compared pairwise, did not reveal polymorphism at the Hind III site within each species. The restriction pattern with respect to Hind III site appears to be species specific in catla, rohu and kalbasu. Further study as suggested by the authors would be useful for stock identification and genetic diversity documentation.

Nuclear DNA RFLP methods were employed to detect inadvertent hybridization in major carp hatcheries when "mixed spawning" of all species Indian major carps is practised (Padhi and Mandal, 1996). It was reported that when southern blots of genomic DNA of the mixed spawned individual offspring digested with EcoR 1 were probed with radiolabeled Xenopus r-DNA the hybrids of catla-mrigal and rohu-mrigal could be detected easily. However, in the case of catla-rohu hybrids caution had to be exercised as the fragment sizes of higher molecular weight diagnostic bands in these species did not vary much. It was also reported that ribosomal DNA fragments were inherited biparentally in catla-rohu, rohu-mrigal and catla-mrigal hybrids. An earlier study on ribosomal DNA RFLP indicated that catla, rohu and mrigal were polymorphic at the EcoR 1 site in ribosomal DNA.

Recently, Majumdar *et al.*, (1997), Basavaraju *et al.*, (1997) and Padhi *et al.*, (1998) have reported on DNA fingerprints in *L. rohita* and *C. catla* by using Bkm 2 C (8) and M 13 multilocus probes. In *L. rohita* and *C. catla*, M 13 gave fewer bands, though the level of polymorphism was higher. This indicates the usefulness of M 13 as an effective DNA fingerprinting probe (Majumdar *et al.*, 1997).

It was also mentioned that although Bkm 2 (8) and M 13 show similar hybridization patterns, they detect different alleles. M 13 gave only a few bands in Indian major carps which were reported to have clustered in 2-4 Kb size ranges when three different enzymes were used to generate such RFLPs, whereas Bkm 2 (8) detects a large number of bands in the size range of 2-20 Kb.

It was further reported that DNA fingerprinting analysis in *L. rohita* and *C. catla* showed similar patterns with the same probe and enzyme combination (Majumdar *et al.*, 1997).

Basavaraju *et al.*, (1997) developed and used four tetranucleotide microsatellite probes to differentiate between broods of catla from the key hatcheries namely, the Tunga Bhadra Dam (TBD), the Bhadra Reservoir Project (BRP) and the Kabini Reservoir Project (KRP) and from out of state wild and hatchery stocks. This analysis was effective in determining the overall levels of genetic variation between strains in terms of number of alleles and average heterozygosity and in detecting differentiation between strains.

These studies also revealed that despite having a common source, the hatchery populations in Karnataka had differentiated from each other as a result of genetic drift or unconscious selection. It was however, reported that there were significant differences in the number of alleles or in the average heterozygosity between the hatchery and wild caught stocks (Basavraju *et al.*, 1997).

Characterization of Mbol satellites in *Cirrhinus mrigala* has been reported by Padhi *et al.*, (1998). These workers claim that they have cloned and characterized a highly reiterated, tandamly repeated and A + T rich Mbol DNA fragment in *Cirrhinus mrigala*, with a monomer size of 266bP. It has been reported that in *Cirrhinus mrigala* the Mbol fragment is species-specific and absent in *C. catla* and *L. rohita*. They have also studied the inheritance of the *Cirrhinus mrigala* specific Mbol fragment in two intergeneric hybrids, *Cirrhinus mrigala* X *L. rohita* and *Cirrhinus mrigala* X *C. catla*, when the radiolabelled32 fragment was hybridized to the Mbol-digested nuclear DNA of *C. catla, Cirrhinus mrigala, L. rohita* and the two hybrids on the southern blot, the *Cirrhinus mrigala* Mbol ladder was found in both the hybrids suggesting that the Mbol satellite in these hybrids was inherited uniparentally (Padhi *et al.*, 1998).

Aneuploidy of a karyotype is due to the loss or gain of one or two chromosones but not due to the loss of an entire chromosone set. Triploids in which all the tree chromosones sets were present were found to be fertile and those which showed chromosome deletion and breakage (aneuploid state) were reported sterile.

Genetic characterization of the founder stocks of *Labeo rohita* from river Ganga, Gomati, Yamuna, Sutlej and river Brahmaputra and also the farm stock from CIFA, which are being used for selective breeding purpose is being carried out. Results of the preliminary investigations have indicated adequate genetic variability (mean heterozygosity 0.047 to 0.079) similar to wild stocks and no genetic contamination was detected (Kuldeep Lal *et al.*, 1998, unpublished).

Until recent past particularly in India aspects like the geographical distribution and biodiversity of fishes were given priority to estimate the extent of distribution and availability or occurrence of different species. However, after realizing the important role of genetics, many a fishery researches too felt the need to exploit the genetic potentials of several commercially important species of fish. To achieve this, information on genetic diversity or divergence, both between and within species is essential. Until recently no attempt has been made with regard to Indian major carps in this direction. However, as discussed in this section, researchers in some of the Asian countries particularly in India have initiated genetic characterization of these carps by applying both biochemical and molecular genetic techniques. Most of these studies are an attempt based on isozyme pattern and some on mtDNA or DNA fingerprinting by using RAPD or RFLP techniques. These preliminary studies with regard to the genetic status of the hatchery and wild stocks, the expression of C-gene pattern in catla, rohu, mrigal and kalbasu, and the pattern of haemoglobin, plasma and transferrin in major carp hybrids and their respective parent species though provided some information, did not indicate any thing which can focus some light on the genetic diversity or divergence either between or within the major carp species.

Much is yet to be done to find out the genetic divergence of these economically most important Indian carps. As mentioned, some methodologies are now available to proceed further in characterization and differentiation, to find out whether there exist different populations/stocks/strains among and within the species and the extent of genetic diversity a variation of these carps. Now the research in this direction has been already initiated in India and is in progress.

2.5 Present status of major carp germplasm resource in relation to environmental changes and habitat modifications

It has been reported that there is considerable depletion in the fish populations of almost all economically important species, including major carps in all the rivers (Das & Barat, 1990). This appears to be mainly due to man made stresses by over fishing, more due to greed for profit making business than simply to make a living by subsistence fishing. Besides, there are also other factors like construction of dams and hydro-power plants etc. which prevent the fish from migrating to their natural breeding and feeding grounds. Pollution of aquatic environment mainly due to heavy industries is the concern of the day. The destruction of fish habitat by alterations of river systems, land development, increased water abstraction etc. is very clear (Jhingran, 1984). Mass mortality of fish in seas and rivers is not very uncommon these days. Consequently, several fish species are feared to be at various stages of threat, such as rare, vulnerable, endangered etc. (National Seminar on Endangered fishes of India, Allahabad, (UP), India, 25-26 April, 1992).

According to Ponniah (1997), rohu, catla and mrigal may not be under threat but they are prone to loss of genetic diversity and variability due to extinction of genetically distinct wild populations, escape and ranching of hatchery seed and competition from exotic carps.

Consequent to massive introduction of cultured fish, displacement of wild stocks has been reported and high levels of introgression of the altered hatchery fish genome into wild stock has also taken place. Escape of farmed fish and breeding with wild populations may result in decreased fitness of wild stock.

Another aspect, which may also pose a problem is the growing trend of replenishing the rivers, lakes and reservoirs with hatchery produced fish seed in many states in India (Mishra and Raman, 1993) and even in the neighbouring countries like Bangladesh. This should be given proper thought as most of the hatcheries produce seed without following any genetic resource management (Eknath and Doyle, 1985) and as such it may be disastrous from a genetic view point. Deviations brought about in the gene pool of survivors to that of the original populations effected by pollution will ultimately be reflected in future populations (FAO, 1980). There is likelyhood of genetic deterioration in populations which suffer drastic depletions. Population size is the single most important factor in sustaining a high level of genetic variation within a population (Frankel and Soule, 1981). Therefore, it is important to restore the size of populations whenever trends of decline are indicated. Das and Barat (1990) have dealt with fish conservation methods that are necessary following habitat degradation. Obviously, the various causative factors inducing stresses on fishes need to be eliminated or reduced as a first step. However, it appears that very little or nothing can be done with regard to alterations of river systems and increased water abstraction when the greater interest of the nation is put ahead of the interest of fish or local groups of fishers or farmers (Das and Barat, 1990).

In situ and *ex-situ* conservation methods have been also often suggested for the restoration of the original gene pool of the species in question. The major advantages of *in situ* conservation according to Das and Barat (1990) are i) continued coevolution in the wild which provides the breeder with a dynamic source of resistance to the changes in the environment is not possible in *ex situ* conservation ii) maintenance of a carp relative *in situ* permits the breeder to study its ecology and to obtain data that can assist in the selection of germplasm that might otherwise be over looked.

In view of the deteriorating aquatic environment, the first necessary step taken should be the conservation of the germplasm resources of at least all the economically important species either in *in situ* (in the natural environment). *Ex-situ* conservation (Cryopreservation and gene banking) can support *in situ* measures, but should not replace them (Pullin *et al*., 1991). This has been also stressed in many seminars/symposia or workshops during the past (Jhingran and Gupta, 1989; Nasar and Haque, 1989) and during the National Seminar on Animal Genetic Resources and their conservation, held at Karnal (Haryana State) India from 22-23 April, 1993.

Conservation of genetic stocks, both wild and cultivated species is important because their genetic diversity has been developed through thousands of years of evolutionary process (Swaminathan, 1984). Selected gene pools or genetic stocks having sufficient genetic diversity and their *in situ* preservation provides the genome the scope of acclimatization with the changes in the environment.

Ex situ method enables the preservation of genotypes in large quantities in an economical way without much management problems. However, this method may also have its disadvantages. For example, in the long run, genome preserved for several decades, when it is revived, may find the prevailing environment hostile . In the case of animals unless both male and female gamates or embryos could be cryopreserved, full advantage of *ex-situ* conservation can not be derived, without further genetic manipulations (Mc Andrew *et al*. 1993).

Both *in situ* and *ex-situ* conservation methods, no doubt have their advantages as well as disadvantages as discussed above. It may be possible to overcome such situations by following rather a cyclical approach through the alternate use of these two methods by intermittently reviving such *ex-situ* preserved germplasm and maintain them again in *in situ* and in the process to expose and acclimatize the genome to the prevailing environment.

The development of molecular genetics offers another novel and great potential method of preserving genetic materials. The genetic material of individuals having the most desirable and heritable traits can be preserved in the form of genomic libraries.

3. HYBRIDIZATION

Hybridization in nature among fish species has been reported by several workers (Hubbs 1955, and Slastenenko 1957). Indian major carps are no exception.

3.1 Hybridization of Indian major carps in nature

The occurrence of fish hybrids in nature has been reported from time to time by several fishery workers. Slastenenko (1957) has given a list of natural hybrids of the world which suggests that species belonging to the cyprinid group are more prone to interbreed than other groups of fishes and Indian major carps belong to this group.

Natural occurrence of both interspecific and intergeneric hybrids of Indian major carps has been reported mostly from reservoirs, formed due to the construction of dams across rivers. These hybrids were also reported from dry "bundhs". Occurrence of rohu-catla hybrid has been reported from the perennial irrigation tanks (reservoirs) in the state of Madhya Pradesh in India. A detailed study of the taxonomic characters of this hybrid has been made by Desai and Rao (1970). Occurrence of catla-rohu hybrids has been reported from Adhartal lake also in Madhya Pradesh. On the basis of the morphological characters and position of the pectoral and anal fins, the parental species of these putative hybrids have been identified as female catla and male rohu (Tripathi *et al.*, 1974). Natural hybrids of similar type have been also reported by Natarajan *et al.* (1976) from the Rehand reservoir in the state of Uttar Pradesh. The impoundment with an area of 46, 538 ha has a mean depth of 25m. The dominant ichthyofauna in the reservoir is *Catla catla* which contributes as much as 98% of the fisheries (Anon, 1974).

The hybrids as reported by the authors possess similar traits but differing traits which reflect the broad spectrum of character variations in hybrids. These natural hybrids as observed by Natarajan *et al.* (1976) have a general appearance of *C.catla* with a conspicuously smaller head than catla but with a deeper body than rohu. The shape of the mouth was more similar to rohu with infringed lips. The hybrid reported from Madhya Pradesh (Desai and Rao, 1970) on the contrary had feebly fringed lower lip and the mouth showed one rudimentary maxillary pair of barbels.

The morphometric traits (ratios) of these natural hybrids more resembled catla in respect of length of head/diameter of eye and length of fish/length of body, whereas, in respect of length of head/height of head ratio, the hybrid resembled rohu, the intermediate traits being the ratio of length of fish/depth of body. The hybrid has also showed intermediate traits in respect of meristic counts of dorsal and pectoral fin rays. Natarajan *et al.* (1976) have given the full details of morphometric and meristic characters of these hybrids.

Occurrence of other carp hybrids like *Labeo fimbriatus* (Bl) x *L. gonius* (Ham.) and *L.calbasu* x *C.catla* in Rangwan reservoir in Uttar Pradesh has been also reported by Prasad (1976). He suggested that these hybrids form good substitutes for *L. gonius* and *L. calbasu* in reservoirs.

Similarly, the hybrids of catla-rohu and rohu-catla would also form a good substitute for rohu in reservoirs. Natarajan *et al.* (1976) mentioned that the studies carried out in respect of the ecology and fisheries of Freshwater reservoirs in Bhavanisagar (Tamil Nadu), Nagarjunasagar (Andhra Pradesh), Govindasagar (Himachal Pradesh), Rihand (Uttar Pradesh) under the all India Co-ordinated Research Project and also D.V.C. reservoir (Jhingran and Natarajan, 1969) and Tungabhadra (David *et al.*, 1969) have shown that the performance of *Labeo rohita* in reservoir ecosystems was not much satisfactory when compared to that of *Catla catla, L. calbasu* and *Cirrhinus mrigala*. The catla-rohu hybrid, having wider spectrum than rohu was suggested to be a good substitute to rohu. *Ceratium* sp. constitutes a very important fraction among phytoplankters in many reservoirs. This ecological niche is not utilized properly. Introduction of Catla-rohu hybrid may help boost the fish production from reservoirs through the proper utilization of the above mentioned niche (Natarajan *et al.,* 1976).

Several causes have been suggested to account for the occurrence of hybrids in nature. Hubbs (1955) holds the view that hybridization in nature is facilitated when there is scarcity of one species and dominance of an allied species in close proximity.

In "bundh" breeding occurrence of natural hybrids may be due to congestion in the spawning ground. Because of the limited space, there is very likelihood of ova of one species getting accidentally fertilized by the sperm of another species. Particularly, in the case of Indian major carps with very compatible genomic structure (Zhang and Reddy, 1991), it is all the more easy and frequent to encounter natural hybrids when bred together in a relatively congregated condition.

3.2 Artificial/experimental hybridization in Indian major carps

In any genetic improvement programme, it is necessary to be able to breed the candidate species artificially under controlled conditions. Soon after the success of induced breeding of the Indian major carps, work on genetic improvement of these carps was initiated in the late fifties starting with simple interspecific and intergeneric hybridization to produce and evaluate their useful traits for aquaculture, since hybridization was regarded to be one of the methods for combining desirable traits of selected species. Accordingly, hybridization work was carried out between different species of a genus (interspecific) and species of different genera (intergeneric) among the Indian major carps and Indian and Chinese major carps, including the common carp (*Cyprinus carpio* variety *communis*, Linnaeus,) to study whether the hybrids acquired any useful traits from their respective parent species, such as smaller head, wider body and more flesh content as in rohu-catla hybrid, pond breeding habit as in *Cyprinus carpio*, or advantageous feeding habits . (Chaudhuri, 1959 and 1973; Alikunhi and Chaudhuri, 1959; Alikunhi *et al.*,1963a and 1972; Naseem Humsa,1971 and 1972. Naseem Hamsa and Alikunhi, 1971; Varghese and Sukumaran, 1971; Ibrahim, 1977; Konda Reddy, 1977; Varghese and Shantharam, 1979; Basavaraju and Varghese, 1980a, 1980b; 1981, 1983; Basavaraju *et al.*, 1989; 1990; 1994 and 1995; Ibrahim *et al.*,1980; Bhowmick *et al.*, 1981; John and Reddy, 1986; Kesavanath *et al.*,1980; Konda Reddy and Varghese, 1980, 1980a, 1980b and 1983; Khan *et al.*, 1989 and 1990; Jana, 1993). These experiments have demonstrated a high level of compatibility among Indian carps for hybridization.

Altogether six interspecific and 13 intergeneric hybrids have been produced among the four species of Indian major carps belonging to the three genera , viz. *Catla, Labeo* and *Cirrhinus*.

Some of the interspecific and intergeneric hybrids of Indian carps and also between Indian carps and common carp (*Cyprinus carpio* var. *Communis*) which were observed to possess some useful traits from aquaculture point of view are described here. Information regarding other hybrids involving Indian major carps is given in the Annex.

3.2.1 Interspecific hybrids

As mentioned above, among the six interspecific hybrids produced and evaluated, characteristics of some useful interspecific hybrids were found (Table 4).

3.2.2 Characteristics of the interspecific hybrids of Indian major carps

Almost all the hybrids, either interspecific or intergeneric, produced among Indian major carps in general exhibited intermediate traits to their parental species.

3.2.2.1 *Labeo rohita* x *Labeo calbasu* (rohu-kalbasu) and *L. calbasu* x *L. rohita* (kalbasu-rohu) hybrids

The percentage of fertilization was very high reaching about 94% and development of eggs was normal. The offspring of both the hybrid crosses were highly viable and their growth rate was superior to their slow growing parent viz. ***L.calbasu*** (Chaudhuri, 1971). Hybrids of both the crosses were found to be fertile and attained maturity within two years. Plates 2 and 3.

3.2.2.2 L. fimbriatus x *L. rohita* (fimbriatus-rohu) hybrid

Basavaraju *et al.* (1990) made a detailed investigation regarding the growth, food conversion ratio and protein efficiency ratio of the hybrid produced between fimbriatus (female) and rohu male. The growth of the hybrid was evaluated in comparison to the parent species. In an 182 day rearing experiment the hybrid was reported to attain 247.8g as compared to 324.5g in rohu and 161.7g in fimbriatus. The growth in the hybrid was 34.75% higher than that of fimbriatus and 23.55 less than rohu. The hybrid also showed the highest survival, being 90% compared to rohu with 86.67% and fimbriatus, 76.67%. In terms of percentage weight gain, specific growth rate and feed utilization, the hybrid performed better than both the parental species.

Plate 2. Interspecific hybrids cross between female *L. rohita* and male *L. calbasu*
a. L. rohita *b.* Hybrid *c. L. calbasu*

Plate 3. Reciprocal interspecific hybrid-cross between female *L.calbasu* and male *L. rohita*
a. L. calbasu *b.* Hybrid *c. L. rohita*

3.2.3 Fertility of interspecific hybrids

Almost all the interspecific hybrids matured within the third year of age (Chaudhuri, 1971 and 1973).

Table 4. List of interspecific hybrids between Indian major carps

FEMALE MALE	HYBRID	COMMENTS
1 *L.rohita* x *L.calbasu*	(rohu-kalbasu)	Hybrid grows faster than Kalbasu and fertile.
2. *L.calbasu* x *L. rohita*	(kalbasu-rohu)	Hybrid grows faster than Kalbasu and fertile.
3. *L.rohita* x *L. bata*	(rohu-bata)	Hybrids exhibited poor hatching and slow growth rate.
4. *L.calbasu* x *L. bata*	(kalbasu- bata)	Hybrids exhibited poor hatching and slow growth rates.
5. *L. gonius* x *L.calbasu*	(gonius-kalbasu)	Hybrids exhibited poor hatching and slow growth rates.
6. *L. fimbriatus* x *L.rohita*	(fimbriatus-rohu)	Hybrids grow faster than fimbriatus and have better feeding efficiency.

3.2.4 Production of F2 generation of interspecific hybrids

The matured kalbasu-rohu hybrid has been bred through hypophyzation and the F2 generation was produced successfully in 1960. As in the case of the F1 progeny, the percentage of fertilization in the F2 progeny was also recorded to be high. The hybrids possessed varying characters which were intermediate between the original parents viz. *Labeo calbasu* and *Labeo rohita* (Chaudhuri, 1971).

3.2.5 Intergeneric hybrids

Of the 13 intergeneric hybrids produced (Table 5) only four hybrids, *L. rohita-C. catla, Cirrhinus mrigala-C.catla, L.rohita-Cirrhinus mrigala, L. fimbriatus-C. catla* and the reciprocal hybrids have been found to possess useful traits in terms of growth (Ibrahim, 1977; Basavaraju and Varghese, 1980 b). The following are the characteristics of these hybrids.

3.2.6 Characteristics of some useful intergeneric hybrids

3.2.6.1 *L.rohita* x *C. catla* (rohu-catla) hybrid

Phenotypically the hybrid exhibits intermediate traits to that of the parents, but tending more towards the maternal parent (L.rohita) with regard to body proportions and fin ray counts. Coloration of the body of the hybrid was near to catla while the shape of the mouth was like rohu. However, its position with infringed lips was terminal and devoid of barbels. The hybrid possessing a smaller head than that of catla and wider body than rohu shows improved traits over the parents. The flesh content of the rohu-catla hybrid, 54-58%, has been also reported to be the highest when compared to rohu with 48-52% and catla with 45% (Bhowmick *et al.*, 1981 and Jana, 1993). Plate 4.

Growth

With regard to growth traits, the hybrid did not differ in growth from its parents in the initial stage. However, in the grow-out stage the growth rate of the hybrid has been reported to vary. Chaudhuri (1973) observed the growth rate of this hybrid was like that of catla, while Chaudhuri *et al.*, 1975 reported, it to be 718-915g in one year duration in carp polyculture experiments, whereas Bhowmick *et al.*, 1981 observed it to be 900-1035g within the same duration. Other workers, (Varghese and Shantharam 1979, Keshavanath *et al.* 1980, Konda Reddy and Varghese 1980, Varghese *et al.*1984, and Somalingam *et al.* 1990) have also reported that the rohu-catla hybrid grows better than rohu.

Plate 4. Intergeneric hybrid-cross between female *L. rohita* and male *C. catla*
a. L. rohita *b.* Hybrid *c. C. catla*

Konda Reddy and Varghese (1980b) conducted experiments on the growth performance of the rohu-catla and catla-rohu hybrids along with the parent species. They observed that the rohu-catla hybrid (female rohu and male catla) showed slightly faster growth than rohu but very much slower than catla.

Keshavanath *et al.* (1980), also evaluated the growth performance of the hybrids produced between *L. rohita* (female) x *C. catla* (male) and the reciprocal hybrids along with their parental species under communal rearing and also independently. These studies have shown that the growth of both the hybrids was inferior to catla and almost equal to rohu. The growth difference was insignificant between the hybrids. However, when reared separately, the growth performance of the hybrids was better than rohu but inferior to catla.

Varghese and Shantharam (1979) made a comparative study on the growth of three intergeneric hybrid of Indian major carps viz., hybrids produced by crossing female rohu x male catla (rohu-catla), catla female x male mrigal (catla-mrigal) and rohu female x mrigal male (rohu-mrigal). After 90 days of rearing, the rohu-catla hybrid showed the fastest growth among the three hybrid types, followed by the catla-mrigal hybrid. Whereas, the rohu-mrigal hybrid attained the lowest growth of all the three. In terms of percentage increase in the growth the rohu-catla and the catla-mrigal have respectively shown 257.6% and 110.9% more over the rohu-mrigal hybrid.

According to the observations of Chaudhuri (1973) among the Indian carp hybrid rohu-catla was the best as it combined the quick growth of catla and small head of rohu. He reported that the growth of this hybrid was faster than rohu and nearly as fast as catla. However, the observations of Konda Reddy and Varghese (1980 b) have shown that this hybrid grew slightly faster than rohu, but its growth when compared to catla was very poor. Somalingam *et al.* (1990) studied the growth of rohu-catla hybrid stocked in three different water bodies, i.e., ponds and small and larger reservoirs. The growth rate of the hybrid in ponds when compared to the parent species was 10-20% less than the male parent catla and 30-125% more than the female parent rohu. However, in small reservoir the growth of the hybrid during the second year of rearing equalled to catla though it was less by 20% and 32% during the first and third years of rearing. At the same time, the hybrid has shown 53%, 128% and 32% higher growth over rohu during the first, second and third year respectively. The authors suggested that the rohu-catla hybrid is a good replacement for rohu to boost the production from the reservoirs.

A more detailed study, conducted at the fish farm of the Central Institute of Freshwater Aquaculture (CIFA), Bhubaneswar in different combinations of the hybrid with the major carp species, i.e. catla, hybrid and mrigal; rohu, catla and hybrid, and mrigal, rohu and hybrid have shown that the rohu-catla hybrid grew better than rohu and mrigal (Jana, 1993).

The morphometric traits of rohu-catla and the reciprocal hybrids have been analyzed by different workers in comparison to the parent species. According to Konda Reddy and Varghese (1980 a) the rohu-catla and catla-rohu hybrids exhibit intermediate traits in respect of the morphometric ratios such as length of fish/length of head, length of head/snout length, length of head/width of mouth and length of fish/length of dorsal fin. In other aspects like length of head/interorbital space, the hybrids tend more towards catla. Rohu-catla hybrid resembled more of catla in the case of the ratios of length of head/diameter of eye, while the reciprocal hybrid catla-rohu tends more towards rohu. The taxonomic characters of these hybrids according to these authors with regard to meristic counts have been observed to be similar, (data not shown).

Bhowmick *et al.* (1981) conducted significance test of rohu-catla hybrid for the fin ray counts and nine body proportion ratios and compared them with the parental species rohu and catla. The values showed that the ratio of body length/head width differ significantly from catla ($P<0.01$), whereas standard length/head length and standard length/head width are significant at 5% level ($P<0.05$). Significant differences were also found for fin ray counts of the pectoral fin ($P<0.01$). These authors have also stated that, although the hybrid appears to be intermediate in general appearance to the parent species, the significance analysis clearly indicates that the hybrid has acquired more of maternal (rohu) traits (data not shown).

Jana (1993), while working for his Ph.D. made a thorough investigation of the rohu-catla and catla-rohu hybrids with regard to their biological and cytological aspects in comparison to their parental species. According to him the morphometric characters of the hybrids and the parent species indicated that the mean values of trunk length and body weight of the rohu-catla hybrid differed significantly ($P<0.001$). It was also observed that head height and body height in trunk length and head width was significantly different ($P<0.01$) from catla.

Jana (1993) mentioned certain key characteristics of the rohu-catla and catla-rohu hybrids which may help to identify these hybrids without much difficulty. In rohu-catla hybrid (rohu female and catla male) the pectoral fins almost reaches the base of the pelvic fin which is a paternal parent trait in the case of catla, where the pectoral fin almost touches the base of the pelvic fin unlike in rohu which leaves a larger gap between tip of pectoral fin when extended towards the base of the pelvic fin. In the case of the reciprocal hybrid, catla female and rohu male (catla-rohu), the pectoral fin when extended, does not touch the base of the pelvic fin and leaves some gap.

3.2.6.2 *C. catla* x *L. rohita* (catla-rohu) hybrid

The compatibility in this reciprocal cross (catla-rohu) was moderately higher as indicated by the percentage of fertilization and hatching, 65-80% and 60-80% respectively. The incubation period for the eggs to hatch out was reported to be from 16.40 to 20.50 hrs after fertilization. The size of 72 hrs hatchling varied, from 6.8 to 7.2 mm.

Konda Reddy and Varghese (1983) produced this catla-rohu intergeneric hybrid and studied the embryonic and larval developments and compared with the parental species. They reported that the developmental stages of the hybrid were quite normal and no abnormalities were noticed and were similar to the parental species.

With regard to the external morphological features, both rohu-catla and catla-rohu hybrids were reported to show similar intermediate traits. Catla has an upturned month with infringed lips while rohu has a ventrally pointed mouth with fringed lips. However, both the hybrids possess a terminal mouth with slightly fringed lower lip. The hybrids also possess a pair of maxillary barbels but shorter and not as prominent as in rohu. Konda Reddy and Varghese (1983) and Jana (1993). Konda Reddy and Varghese (1980b) also reported that the mouth parts and gill rakers of both the hybrids are very similar, indicating the same food habits. The observation of Alikunhi *et al.* (1971) have shown the hybrid to grow better than even catla under monoculture as well as when grown with catla. Similarly, Varghese and Sukumaran (1971) also observed the hybrid to grow better than even both the parents. This is in contrary to what was reported by Konda Reddy and Varghese (1980 b). These workers reported that the catla-rohu hybrid (catla female-rohu male) grew very much slower than catla and slightly slower than rohu. Rohu registered 25.2%

Plate 5. Reciprocal intergeneric hybrid-cross between female *C.catla* and male *L. rohita*
a. C. catla *b.* Hybrid *c. L. rohita*

more than the hybrid while catla registered 108.9% more. These authors concluded that this hybrid is much inferior to catla and slightly slower than rohu (Plate 5).

The reasons for this, according to Konda Reddy and Varghese (1980 b) seem to be that the food habits of these hybrids as reported by Chaudhuri (1973) were basically detritophagic in pond environment and more like rohu. So there appears to be food competition between the hybrids and rohu. Natarajan *et al*. (1976) also reported that in reservoir environment the hybrid was detritophyto-planktophagic and has shown active inclination towards detritus and phytoplankton in preference to zooplankton, which constitutes the chief food of catla.

3.2.6.3 <u>*C. catla* x *Cirrhinus mrigala* (catla x mrigal) hybrid</u>

This intergeneric hybrid cross exhibited relatively high compatibility as evidenced by the high rate of fertilization, over 97%. The fully swollen and developing zygotes measured 5.8 mm while the hatchling measured over 4.0 mm, which is much smaller than the catla-rohu hybrid.

Ibrahim (1977) made a thorough investigation of the hybrids produced between *C. catla* female and *Cirrhinus mrigala* male and the reciprocal hybrids and also the hybrid between *L. rohita* female and C. mrigala male. According to him the catla-mrigal hybrid had an elongated body with a terminal but slightly ventrally placed mouth. The upper lip was slightly thick and the head more like catla. Pectoral, pelvic and anal fins were slightly pinkish but not as much as in mrigal. The hybrid was devoid of barbels. The snout was slightly blunt tending towards the paternal parent, the mrigal. Though not as pronounced as in catla, the hybrid had a slight hump. The body color was more like mrigal. Diameter of the eye was larger in mrigal than in catla. But in the hybrid the eye was much larger than the parents. Diameter of the eye in relation to head length was 6.89 mm in catla, 5.57 mm in mrigal and 4.129 mm in the hybrid. From the view point of the morphometric characters, the hybrid differed from the parental forms with regard to fin ray counts, dorsal, pectoral and pelvic fin rays, and tended more towards the maternal parent, the catla.

<u>Food and feeding</u>

Catla, a plankton and surface feeder has 245 gill rakers, which are slender, elongated and suitable for filtering, whereas in mrigal, which is a bottom feeder as also mentioned earlier, has only over 60 short and slender gill rakers which are placed widely. The food niches and food habits of the two parental forms are quite different. The hybrid possesses 91 gill rakers, which is somewhat intermediate but nearer to the paternal parent, the mrigal. Thus the hybrid though gained a little over mrigal in the filtering mechanism, did not gain over the mother.

Again, the hybrid with alimentary canal length of 6.15 times the total length, resembles the maternal parent, catla in which it was 6.21 times. However, in mrigal it was much longer, being 18 times, the total length (Ibrahim 1977). Sex ratio of the hybrids was 66.6% females and 33.4% males as reported by Ibrahim (1977).

Growth

Ibrahim (1997) made a detailed study on the growth of catla x mrigal hybrid and the reciprocal hybrids and compared them with the parental species in communal ponds. According to him the catla x mrigal hybrid attained an average growth of 107.7mm/10.3g in 11 months while the parental forms catla and mrigal attained 160.3mm/45.8g and 174.0mm/55.5g respectively. The growth of the hybrid, thus was not as good as the parental forms.

3.2.6.4 *Cirrhinus mrigala* x *C. catla* (mrigal-catla) hybrid

Unlike in the reciprocal hybrid, this hybrid possessed an up-turned mouth with the upper lip slightly over hanging. Lower lip not fringed and without barbels. The hybrid was reported to resemble the catla (father) with regard to the head and dorsal profile. The hump of catla was also clear in the hybrid. (Ibrahim, 1977).

The hybrid was found to be intermediate to the parents with regard to the anterior portion of the body in relation to its total length. The depth of the body in the hybrid was greater than that of the mrigal as the former has inherited the hump of catla. When compared to catla and rohu, mrigal has a minimum hump. The increase in the body depth due to the hump can be regarded as a positive trait in the hybrid as in the case of the rohu-catla hybrid. The hybrid had an increased size of head in relation to total length (4.58), compared to the mrigal (3.35) and smaller than in catla (3.33). The eye diameter was relatively larger than in catla but not as large as in mrigal. The significant morphometric characters are thus the depth of body and size of head of the hybrid which are positive gains (Ibrahim, 1977).

The fin ray counts in the hybrid with regard to the dorsal and pectoral fins, have shown differences from those of the parents.

The pharyngeal complex in the hybrid with 144 gill rakers has shown significant deviation from both the parents which, as mentioned earlier in mrigal was over 60 and 245 in catla. The increase in the number of gill rakers according to Ibrahim, (1977) is an indication of the hybrid adoption towards plankton food habit as in catla.

Growth

The growth performance of this hybrid (mrigal x catla) according to Ibrahim (1977) was better than either parents. The hybrid grew to an average size of 221.2mm/105g in 11 months compared to growth of catla and mrigal being 160mm/45.8g and 174mm/55.5g respectively.

3.2.6.5 *L.rohita* x *Cirrhinus mrigala* (rohu-mrigal) hybrid

The experiments of Ibrahim (1977) have shown that the hybrid cross had a very high percentage of fertilization (97.9%) indicating the highly compatible nature between the genomes of these two intergeneric species. The incubation period for the eggs to hatch was 15 hours and 30 minutes. The hatchlings, immediately after hatchling measured 3.67mm. By about 92 hours the yolk sack in the hatchling (spawn) was completely absorbed. The hybrid had a ventrally turned mouth with the upper lip over hanging. A maxillary pair of barbels was reported to be present. The lower lip was not fringed as in the case of rohu (maternal parent). The hybrid possessed slightly greenish color on the dorsal profile as in the case of rohu. However, the ventral side with silvery color resembled the paternal parent (mrigal). Also, coloration of pectoral, pelvic and caudal fins resembled that of mrigal. The snout was slightly pointed, resembling rohu.

The head length and body height of the hybrid were longer than the father. However, the hybrid showed lower gill-raker length than both the parental . A positive trait in the hybrid was that it possessed a larger number of gill rakers than both the parents. This according to Ibrahim (1977) is a probable indication of the hybrid tending towards consumption of more planktonic food than either of the parents thus providing the hybrid a wider spectrum in feeding. The sex ratio among the hybrids according to Ibrahim (1977) was 35% female, 65% male and some maturing specimens of hybrids were noticed during the third year.

Growth

The growth of this rohu-mrigal hybrid was also good attaining an average size of 242.8mm/143g in 11 months. Among these three intergeneric hybrids, catla-mrigal, mrigal-catla and rohu-mrigal, whose growth performance was evaluated in communal pond along with the parental forms, catla, mrigal and rohu, the mrigal-catla and the rohu-mrigal hybrids have shown a better overall growth performance than the parental forms (Ibrahim,1977). However, the author mentioned that the growth recorded in this evaluation should not be taken as a complete manifestation of the growth potential as the fishes were confined in a small nursery, which is not considered ideal for maximum growth assessment of fishes. However, the trend observed can be considered significant.

The overall analysis of morphometric and meristic characters of all these hybrids interestingly indicated that the hybrids showed greater divergence from parental forms in characters like body height, head length, eye diameter, gill rakers number and size, and also in the length of the alimentary canal (Ibrahim, 1977).

In the *Cirrhinus mrigala* (mrigal) x *C. catla* (catla), the body height had a gain of 15.8% against mrigal (mother) while it had a loss of 16.5% over catla (father). In head size the hybrid gained 19.9% over mrigal and a loss of 27.6% was observed over catla. In the length of gill rakers the hybrid gained 62.9% over mrigal and there was a loss of 63.3% over catla. Again in eye diameter the hybrid indicated a loss of 1.4% and a gain of 17.6% over the maternal and paternal parental forms respectively.

Considering the other characters, this hybrid has a wider spectrum of feeding in converting more of column-planktonic and bottom feeds. This, possibly could be the reason for a relatively better growth of the hybrid over the parental forms (Ibrahim, 1977).

Basavaraju and Varghese (1980b) studied the growth rates of the *L. rohita* (female) x *Cirrhinus mrigala* (male) and the reciprocal hybrids and compared them with that of the parental species which were also reared along with the hybrids. This study has shown that both the parental species, *L. rohita* and *Cirrhinus mrigala* grew faster than both the hybrids. Again, among the two hybrids the hybrid between *Cirrhinus mrigala* female x *L. rohita* male grew faster than the reciprocal *L. rohita* female x *C. mrigal* male hybrid. Thus, the average weights attained by rohu, mrigal and the reciprocal hybrids, mrigal-rohu and rohu-mrigal were 151.52, 92.37, 72.16 and 61.0g respectively. Whereas the observations of Ibrahim (1977) have shown that growth rate of the rohu (female) x mrigal (male) was better than the parent species.

3.2.6.6 <u>*L. fimbriatus* x *C. catla* (fimbriatus-catla) and *C. catla* x *L. fimbriatus* (catla-fimbriatus) hybrids</u>

Basavaraju *et al*. (1994) studied the morphometric and meristic characters of the hybrid between *L. fimbriatus* female and *C. catla* male in comparison with the parental species. These authors stated that it was difficult to distinguish these reciprocal hybrids.

The fimbriatus-catla hybrid also possessed intermediate characters to parental species in body profile, coloration of fins and structure of lips. Body was deep with a smaller head than in the paternal parent catla. The Mouth was subterminal with a moderately fringed lower lip.

With regard to the morphometric characters, the hybrid resembled catla (male parent) in total length/standard length, length of head/eye diameter and length of head/height of head. Its resemblance with the female parent (fimbriatus) was only in respect of head length/maximum head width. The intermediate characters were total length/fork length, total length/pro-dorsal length and total length/least height of caudal peduncle (Basavaraju *et al.,* 1994).

Growth

Evaluation of the hybrid between female *C. catla* and male *L. fimbriatus* (catla-fimbriatus) and the reciprocal hybrid i.e. female *L. fimbriatus* and male *C. catla* (fimbriatus-catla) has been done by Basavaraju *et al.* (1995). These workers reported that the growth of hybrids was much faster than the slow-growing parent, fimbriatus and less than catla when assessed under similar conditions. On conclusion the growth rates recorded were, catla 64.2 fimbriatus 24g, hybrids fimbriatus catla 57.8g and catla -fimbriatus 39.8g.

The growth increment attained by fimbriatus-catla was 153.4 and 45.0% more than fimbriatus and the reciprocal hybrid respectively and not much different from catla. As in the case of the rohu-catla hybrids, these hybrids also exhibit intermediate traits with desirable qualities such as smaller head in the fimbriatus and deeper body as in catla. The meat yield in the hybrids (62.1% in fimbriatus-catla and 63% in catla-fimbriatus) was higher than in the parents (52.5% in catla and 61.9% in fimbriatus), as in the case of rohu-catla hybrid.

Table 5. List of intergeneric hybrids between Indian major carps

FEMALE MALE	HYBRID	COMMENTS
1. *C. catla* x *L.rohita*	(catla-rohu)	Growth rate of hybridsvariable, grow better than both the parents (Verghese and Sukumaran, 1971), grow slower than both the parents (Konda Reddy & Verghese, 1980), fertile hybrids.
2. *L.rohita* x *C. catla*	(rohu-catla)	Hybrids grow faster than rohu.The meat yield in hybrids is higher than in parents. The hybrids are fertile.
3. *C. catla* x *C. mrigala*	(catla-mrigal)	Hybrids grow slower than the parents and are fertile.
4. *C. mrigala* x *C. catla*	(mrigal-catla)	Hybrids grow better than either parents and are fertile.
5. *L.rohita* x *C. mrigala*	(rohu-mrigal)	Growth rate of hybridsvariable, grow better than the parent species (Ibrahim, 1977), grow slower than both the parents (Basavaraju and Verghese, 1980b). The hybrids are fertile.
6. *C. mrigala* x *L.rohita*	(mrigal- rohu)	Hybrids grow slower than both the parents and are fertile.
7. *C. catla* x *L.calbasu*	(catla-kalbasu)	Hybrids grow faster than kalbasu and they have broader feeding spectrum & adaptability.
8. *L. calbasu* x *C. catla*	(kalbasu-catla)	Hybrids exhibited low viability, need further study with regard to survival & performance.
9. *L.calbasu* x *C. mrigala*	(kalbasu- mrigal)	Fertilization in the hybrid cross was normal (80-90%) with 60% hatching rate. The hybrids are fertile.
10. *L.rohita* x *C. reba*	(rohu-reba)	Hybrids exihibit low viability.
11. *L.calbasu* x *C. reba*	(kalbasu-reba)	Hybrids exihibit low viability.
12. *C. catla* x *L. fimbriatus*	(catla-fimbriatus)	Hybrids grow much faster than *L. fimbriatus*. The meat yield is higher than the parents.
13. *L. fimbriatus* x *C. catla*	(fimbriatus-catla)	Hybrids grow much faster than *L. fimbriatus*. The meat yield is higher than the parents.

3.2.7 Intergeneric hybrids between Indian major carps and common carp (*Cyprinus carpio* var. *communis*)

The hybrids between Indian major carps and common carp (*Cyprinus carpio var. communis*), involving the crosses between common carp male x rohu female, common carp female x males of rohu, catla and mrigal in each cross, respectively have on the whole exhibited normal fertilization and moderate viability, though in certain cases, the crosses involving females of Indian carps with males of common carp, low viability has been exhibited in the hybrids. Details of these intergeneric hybrid crosses are shown in Table 6.

Table 6. Intergeneric hybrids between Indian major carps and Common carp

FEMALE MALE	HYBRID	COMMENTS
1. *L.rohita* x *C. carpio*	(rohu-common carp)	Hybrids exhibited poor survival and are sterile.
2. *C. mrigala* x *C. carpio*	(mrigal-common carp)	Poor survival. The hybrid tends to resemble *C. carpio* and are sterile.
3. *C. carpio* x *L.rohita*	(common carp-rohu)	Higher survival rate than the reciprocal-cross and dominated by mostly *C. carpio* traits. The hybrids are sterile (aneuploid), grow better under monoculture system.
4. *C. carpio* x *C. catla*	(common carp-catla)	Higher survival rate than the reciprocal-cross and dominated by mostly *C. carpio* traits. The hybrids are sterile (aneuploid), grow better under monoculture system.
5. *C. carpio* x *C. mrigala*	(common carp-mrigal)	Hybrids are inferior to both the parents in growth and are sterile.

3.2.7.1 Characteristics of hybrids produced between Indian major carps and common carp *(Cyprinus carpio var. communis)*

The characteristics of these intergeneric hybrids also more or less exhibited intermediate traits, though some traits were characteristic of the maternal parent and in others of the paternal parent.

3.2.7.2 *L. rohita* x *C.carpio* (rohu-common carp) hybrid

Though the percentage of fertilization appeared to be good, the development in many of the eggs ceased during the incubation period. The bulk of the embryos died during and after hatching and only a few hatchlings could grow up to the adult stage which did not show any signs of maturity.

Morphologically, the hybrid appeared to be intermediate between the parents but statistical analysis of various parameters showed it to be closer to the paternal parent (common carp).

3.2.7.3 *C. carpio* x *L. rohita* (common carp-rohu) hybrid

Khan *et al.*, 1990 made a detailed study of this hybrid cross between common carp female and *L.rohita* male. The fertilization rate was as good as in pure common carp crosses. Hatching process was completed within 46.30 hrs at water temperature 27-29.5^0 C .

The dorsally convex and elongated body of the hybrid was broader than that of rohu. In other morphological features the hybrid had a terminal mouth which was also protractile as in common carp. The lips were not fringed but one pair of maxillary and another pair of rostral barbels were present. Regarding the fins, the dorsal fin though long, but was shorter than in common carp. Serrations of the spiny dorsal and anal fin rays which are very conspicuous in common carp, were very feeble in the hybrid. The hybrid resembled the common carp in prepectoral distance, the lateral line position, color of the body. However, the caudal fin in the hybrid was deeply forked with 24 rays. With regard to size of scales, the hybrid was intermediate. The color of fins was also intermediate.

The characteristic big belly of common carp was absent in this hybrid, as in the case of mrigal-common carp hybrid, and the shape of the abdomen resembles that of rohu, the paternal parent. The size of the head in the hybrid was intermediate being smaller than that of common carp but bigger than in rohu. On the whole the hybrid had more maternal traits. Plate 6.

Plate 6. Intergeneric hybrid-cross between female *C. carpio* and male *L. rohita*
a. C. carpio *b.* Hybrid *c. L. rohita*

The survival and growth of common carp-rohu hybrid and common carp under monoculture system were 91.45g and 73% and 76.5g and 71% respectively during a period of 180 days. However, when the hybrid was reared along with the parents, common carp and rohu in equal ratios, the average growth performance and survival were 84.3g and 68.5%, 237g and 70% and 216.7g and 88% respectively. Thus the hybrid has shown slower growth and lower survival than both the parents, when grown together.

Regarding food habits, these hybrids are bottom feeders as the maternal parent and their food items consisted of semidigested crustacean appendages, bottom detritus and sand particles. Diatoms like *Amphora, Navicula* and filamentus alga were also observed in the food. The dress value of the hybrid had a higher percentage of flesh (58%) than that of common carp (47%), but less than in rohu (60.3%).

Regarding fertility, the hybrid did not exhibit any signs of maturity after two to four years of age. However, rudiments of testes were observed in some individuals.

3.2.7.4 <u>*C. carpio* x *C. catla* (common carp-catla) hybrid</u>

The cross of common carp female with male of *C. catla* showed a relatively high percentage of fertilization, ranging from 65-80%, but in this case also about 35-45% mortality occurred during embryonic development, as in the case of the mrigal-common carp hybrid. Most of the hatchlings were deformed with a curve in the caudal region and died 72 hrs after hatching.

The hybrid possessed a dorsally convex and more elongated body than common carp. A slight hump as in the case of common carp was also present. As in the case of the common carp rohu hybrid, the characteristic belly of the common carp was lacking in this hybrid too. The mouth of the hybrid was terminal and shorter than that of common carp. Plate 7.

Plate 7. Intergeneric hybrid-cross between female *C. carpio* and male *C. catla*
a. C. carpio *b*. Hybrid (sterile) *c. C. catla*

The hybrid attained an average weight of 21.0g in polyculture system and 71g under monoculture system (when reared alone) within a period of two months (Khan *et al*., 1990).

Thus, over three decades of research on hybridization among different species of Indian major carps and their evaluation has shown that these major carps though possessing distinct morphological characters, genera, they are highly compatible and able to produce viable and fertile hybrid progenies. Matured interspecific and intergeneric hybrids could be induced to produce F2 progeny or back-cross and triple cross hybrids.

The growth exhibited by most of these hybrids was intermediate i.e. better than the slow growing parent as in the case of the interspecific hybrid between *L. rohita* and *L. calbasu* or the intergeneric hybrids between *L.rohita* and *C. catla* or *C. catla* and *Cirrhinus mrigala* etc.

The gain in the growth in these hybrids over their slow growing parents appears to be due to the inheritance of this trait from the fast growing parent, rohu in the case of *L. rohita-L. calbasu* hybrid and catla in the case of *L.rohita-C. catla* hybrid and so on. Chevassus (1983) also expressed the opinion that the intermediate nature of the growth in the hybrids, may be due to the partial transmission of this trait to the hybrid by the fast growing parent.

The growth of some hybrids like the rohu-catla or mrigal-catla or with their reciprocal cross appears to be due to the available food in the medium and also could be due to the variations that probably existed in the rearing medium under different agroclimatic zones.

With regard to the hybrid crosses between Indian major carps and Chinese grass carp, silver carp and big head carp, no compatibility was observed in any of the hybrid progenies as almost all of them died either during the embryonic development or soon after hatching, maximum within a week, except some stray hybrids produced between *C. catla* and *H. molitrix* (silver carp).

Again, relatively better survival of the hybrid progeny was observed in the crosses between female common carp x males of Indian major carps, viz., *C. catla, L. rohita* and *Cirrhinus mrigala* compared to the reciprocal hybrid crosses particularly between female *L. rohita* and male common carp. These reciprocal hybrids followed the maternal parent in most of the embryonic developmental events particularly, duration of incubation, commencement of cleavage etc. However, the reason the viability of hybrid progeny was poor in crosses where Indian major carps were the maternal parents appears to be due to the haploid nature of most of those hybrid progenies as indicated by their curved caudal region probably caused due to the relatively larger size of the head of the penetrating sperm of common carp (with n=50-51 chromosomes) through the smaller micropyle of the major carp egg (with n=25 chromosomes). Thus often preventing the genome input from the male and thereby proper fertilization.

However, further investigation may be necessary, especially on hybrids between Indian major carps and minor carps and between Indian major carps and common carp.

It is essential to prevent indiscriminate hybridization by checking the mixed spawning of Indian carps. These carps, as has been observed, are highly compatible to interbreed. Also, since almost all interspecific and intergeneric hybrids were found fertile, we should be all the more cautious to check their cross-breeding which may cause serious damage to the original gene pool through genetic introgression. (Padhi and Mandal, 1994, Mishra *et al*., 1998).

Some of these hybrids may be very useful for stocking reservoirs in India, where they are observed to thrive well and contribute substantially to the production (Natarajan, *et al*., 1976).

3.3 Evaluation of meat of hybrids

Basavaraju and Varghese (1980) have made some study on organoleptic evaluation of the meat of rohu-mrigal and mrigal-rohu and compared it to that of the parental species, rohu and mrigal. The test was done for both raw and cooked meat. The attributes tested were color gloss of skin, odor of flesh, color of flesh, and texture for raw fish and color, texture, taste and flavor for cooked meat. Analysis of variance (Krammer and Twigg, 1970, Basavaraju and Varghese 1980) showed that the uncooked flesh of hybrids was found to be inferior in quality to that of the parental species, while no significant difference was found with regard to the cooked flesh.

3.4 Mixed spawning and hybridization

After the development of induced breeding technique for Indian major carps, production of pure seed of these carps became possible. Dependence of the fish culturist for their seed from riverine sources gradually decreased for many reasons. After the advent of the intensive carp culture system, more and more farmers started taking up fish culture. Consequently there has been a growing demand for the seed of major carps. Based on the demand of fish farmers, the seed producers that used Chinese-type hatcheries produced the seed of all three major carps, catla, rohu and mrigal, by breeding them together in a common bedding pool through the administration of hormones. This practice is mostly followed in the state of West Bengal where 70% of the carp seed requirement of the country is met (Padhi and Mandal, 1994). This kind of seed production is generally resorted to by the seed producer, not only to supply seed of all the three species simultaneously in the required quantities, but also to save time as the breeding span of these carps lasts usually for only one and a half to two months. Thus the seed producer's aim is to supply the seed as per the demand and earn as much money as possible, without realizing that there may be genetic consequences of the practice.

Mishra, *et al*. (1998) have assessed the percentage of hybridization taking place in such mixed spawning practices. According to their observations the rate of what they termed as spontaneous/inadvertent hybridization (intergeneric) ranged from 7.24 to 9.24% in a fish seed hatchery, they studied in the state of Uttar Pradesh. The hybrids were identified as catla-mrigal and, rohu-mrigal. The growth performance of these hybrids as observed by Mishra, *et al*. (1998) was that the hybrids in general were found to possess significantly higher lengths and weights than their parent species. The growth of catla-rohu was found to be highest among all. The rohu-mrigal hybrids performed better than their parents. Performance of catla-mrigal hybrids was better than mrigal, but not catla.

Though all the three species bred together (mixed spawning), the incidence of hybridization (7.24 to 9.24%) is not relatively very high in spite of the compatibility of these species to hybridize. The probable reasons for this could be that Indian carps exhibit courting behaviour while spawning where the female is usually followed (chased) by several, males very closely. Here, males might be following their like species during the chase, ultimately resulting in the release of eggs by the female and simultaneously release of milt by the males.

It was also reported that the male coils round the abdomen of the female and releases milt when she releases the eggs The spontaneous/inadvertent hybridization which is reported to take place during mixed spawn may be due to accidental fertilization of the eggs of one species by the sperm of another species which might have been released in close proximity.

Species specific courtship behaviour and generally species recognition keep the accidental level of hybridization low. However, the presence of even a few number of hybrids among the lot of other pure species (Indian major carps) may be detrimental to the genetic quality in the long run.

ANNEX

Information on other hybrids

I. Interspecific hybrids of Indian major carps

i) *L.rohita* x *L. bata* (rohu-bata) , *L.calbasu* x *L.bata* (kalbasu-bata) and *L. gonius* x *L. calbasu* (gonius-kalbasu) hybrids.

As in the case of the kalbasu-rohu and the reciprocal hybrids, the percentage of fertilization and embryonic development of these hybrid offsprings were normal, although the percentage of hatching was reported to be poor. Rohu-bata and kalbasu-bata hybrids did not show much promise, though the latter hybrid exhibited good initial growth.

II. Intergeneric hybrids of Indian major carps

i) *Cirrhinus mrigala* x *L. rohita* (mrigal-rohu)

A detailed, account of embryonic developments of mrigal-rohu hybrid, produced between female mrigal and male of rohu has been given by Basavaraju and Varghese, (1981). Unlike the case of the reciprocal hybrid, rohu-catla in which the fully swollen eggs of hybrid were bigger than those of the parents (Konda Reddy, 1977), in this mrigal-rohu hybrid the eggs were intermediate in size. Contrary to this, the commencement of cleavage in catla-rohu and rohu-catla hybrids as reported by Konda Reddy (1977) was intermediate, whereas in the case of the mrigal-rohu hybrid it was observed to commence later than the parent species (Basavaraju and Varghese, 1981).

The development of the mrigal-rohu, hybrid, in conclusion was normal and similar to that of the parent species. With regard to the morphological features, the hybrid exhibited intermediate traits between the parents. (Basavaraju and Varghese, 1981).

ii) *C. catla* x *L. calbasu* (catla-kalbasu) hybrid

Morphologically the hybrid has a greater resemblance with kalbasu in general appearance, indicating paternal dominance. The hybrid had a smaller head than catla and body girth greater than kalbasu. The quantity of flesh in the hybrid was also greater than in both the parents. The hybrid possessed a terminal mouth with thick lips. The feeding habits of the hybrid seem to be broadly varied as evidenced by the gut contents, comprising crustaceans, chironomid larvae, rotifers, diatoms, other algal forms, sand and mud. The hybrid also relishes artificial feed which was also observed in the gut in various stages of digestion. Thus the hybrid has a broader feeding spectrum and higher degree of adaptability for different food items from various niches of the pond.

The hybrid showed 66% survival at fry stage and at fingerlings, survival at the end of one year rearing in carp polyculture system was 40-80%. It attained a size of 330 mm/425 g as compared to 360 mm/680 g for catla. The growth of the hybrid under culture conditions was faster than kalbasu. Bhowmick *et al.* (1987) studied this hybrid in detail with regard to percentage of fertilization, growth, food and feeding habits, maturity and fecundity etc.. Prasad, 1976 reported the occurrence of *L. calbasu* and *C. catla* hybrid from the Rangwan Reservoir from the state of Uttar Pradesh in India (a natural hybrid).

iii) *L. calbasu* x *C. catla* (kalbasu-catla) hybrid

This cross between female *L. calbasu* and male *C. catla* did not yield any appreciable percentage of viable progeny. The development of eggs after insemination was abnormal and only about 20% of them hatched out, of which only a small percentage survived. However, further experimentation may be needed to arrive at definite conclusions.

iv) *L. calbasu* x *Cirrhinus mrigala* (kalbasu-mrigal) hybrid

This hybrid cross between female *L. calbasu* and male *Cirrhinus mrigala* showed normal fertilization with a 60% hatching rate. The hybrid had a slightly fringed lower lip and two pairs of prominent black barbels resembling that of maternal parent. Body color of the hybrid was intermediate between the parents species. No further details are available regarding the growth performance. However, the hybrids have been reported to attain maturity in 2-3 years.

v) *L. rohita* x *C. reba* (Rohu-reba) and *L. calbasu* x *C .reba* (kalbasu-reba) hybrids

These intergeneric hybrid crosses between female *L. rohita* and male *C. reba* did not result in any viable hybrids. Only about 20% of the hybrid zygotes hatched out but all the hatchlings died on the third day after hatching. Naseem Hamsa and Alikunhi (1971) also attempted to upgrade *C. rebe* by crossing it with the males of rohu, mrigal and grass carp. The hatchlings in all the three crosses were observed to be longer than those of *C.reba*. However, all the hatchling died within two days.

III. Intergeneric hybrids among Indian and Chinese carp

Hybridization was also attempted between, the species of the three genera, viz., *catla, labeo* and *cirrhinus* of Indian major carps and the three species of Chinese carps belonging to the three genera viz., *Ctenopharyngodon, Hypophthalmichthys* and *Aristichthys*. In all, nine intergeneric hybrids have been produced between Indian and Chinese major carps. Many of these hybrid progeny did not survive beyond one week excepting the catla-silver carp hybrids. However high mortality of the hybrids at embryo and larval stages was observed in catla-silver carp hybrid too, up to the fourth day. Many of the hatchlings were deformed particularly at the caudal peduncle region. In one experiment, only three hybrids could be reared up to three months, while in another only 16 surviving hybrids could be reared for a period of over two months and they grew to a size of about 57.7mm within this period (Ibrahim *et al*., 1980).

The percentage of fertilization was reported to be very high, ranging from 82-100%. (Ibrahim *et al*., 1980). Morphologically, the hybrid had a comparable body growth to that of catla. The size of the head and scales, including the body color were similar to silver carp.

In the hybrid progeny resulting from the mrigal x grass carp cross, the complete process of embryonic development including cleavage, segmentation etc. was quite normal and the hatching rate was over 90% . However, the larvae gradually became abnormal with a bent body and died within eight days after hatching (Ibrahim *et al*., 1980).

List of intergeneric hybrids between Indian and Chinese carp

FEMALE MALE	HYBRID	COMMENTS
1. L.rohita x *Hypophthalmichthys molitrix* (silver carp)	rohu-silver carp	Hybrids are not viable beyond larval stage
2. C. catla x *H. molitrix*	catla-silver carp	Few hybrids survived only over two months
3. H. molitrix x *C. catla*	siver carp-catla	Few hybrids survived only over two months
4. L. rohita x *Ctenopharyngodon idella (grass carp)*	rohu-grass carp	Hybrids are not viable beyond larval stage
5. C. mrigala x *C. idella*	mrigal-grass carp	Hybrids are not viable beyond larval stage
6. C. idella x *C. catla*	grass carp- catla	Hybrids are not viable beyond larval stage
7. C. idella x *L. rohita*	grass carp- rohu	Hybrids are not viable beyond larval stage
8. Aristichthys nobilis x *C. catla* (bighead)	bighead-catla	Hybrids are not viable beyond larval stage
9. A. nobilis x *L. rohita*	bighead-rohu	Hybrids are not viable beyond larval stage

IV. Intergeneric hybrids between *Cirrhinus mrigala* and *C. carpio* and thereciprocal cross

i) *Cirrhinus mrigala* x *C. carpio* (Mrigal-Common carp) hybrid

Kowtal and Gupta (1985) produced and evaluated intergeneric hybrid between *Cirrhinus mrigala* female and *C. carpio* male. These authors have reported a high rate of fertilization. However, about 40% mortality was recorded during embryonic and larval development. Only a few hybrids could be reared to adult size, i.e. to 242 mm/ 120 g after 14 months rearing in nursery ponds.

The hybrid had a greater body girth than the maternal parent (mrigal) and a smaller head than in common carp. Though the outline of the mouth was similar to that of mrigal, it was protractile as in the paternal parent (common carp). The hybrid resembled common carp in having an elongated dorsal fin, moderately forked caudal fin and in scale and color pattern.

ii) *C. carpio* x *Cirrhinus mrigala* (common carp-mrigal) hybrid

As in the case of the common carp-rohu hybrid, in this cross the rate of fertilization was similar to pure common carp crosses. In embryonic and larval development the hybrid followed more or less the pattern of the maternal parent (common carp). Hatching commenced from 23.00 hrs after fertilization and was completed by 44-49 hrs. at 27-29 C.

Like the common carp-catla hybrid, this hybrid possessed a dorsally convex body with a slight hump as in common carp, but without the characteristic big belly of the mrigal. However, the body was more elongated and broader than mrigal. The terminal mouth of the hybrid was not protractile as in the case of common carp. Barbels, one pair of maxillary and the other mandibular were present. All this, the shape of snout (blunt) and the eye located in the anterior portion of the head, made the hybrid similar to the common carp. The traits resembling mrigal were the elongated body without big belly and non-protractile mouth and the pattern of lateral line.

The variable color of the hybrid, back greenish, sides golden yellow with a whitish ventral surface and greyish dorsal and caudal is an intermediate in appearances, at the same time the hybrid exhibits traits, which tend more towards the maternal parent, the common carp.

Regarding growth, the common carp-mrigal hybrid attained 48.2g under monoculture system, but it grew to only 10.2g in polyculture during the span of 110 days rearing. However, the parent species common carp and mrigal registered a weight of 107.4g and 110.8g respectively in the same polyculture system under similar rearing conditions. Thus the growth performance of this hybrid was slightly better than only mrigal (paternal parent), when the former was reared under monoculture (Khan *et al*., 1990).

V. Intraspecific hybridization in rohu (*L. rohita*)

Interspecific hybridization was also carried out among the members of stocks from different rivers which was termed as complete diallel crosses. (Gjarda *et al*., 1998). One with the stocks of River Ganga, Local (farm) and River Yamuna (Diallel-1) and the other with the stocks of River Brahmaputra, Local and River Sutlej (Diallel-2). The milt from each sire of each stock (5 sires/stock) was used to fertilize an equal amount of eggs from the females of the same or a different stock (8-10 females/stock).

The results of the two diallel-crosses indicated low average heterosis effects for both harvest body weight and survival. For diallel-2, significant negative average heterosis for body weight was observed in both mono and polyculture production systems. The results thus suggest that systematic crossing of rohu stocks may not yield any progeny with positive heterosis.

No. of crosses (Diallele 1)

Female		Male	Comments
1. Ganga	X	Ganga (GG)	Normal/control cross
2. Ganga	X	Yamuna (GY)	No heterosis exhibited
3. Ganga	X	Local (GL)	No heterosis exhibited
4. Yamuna	X	Ganga (YG)	No heterosis exhibited
5. Yamuna	X	Yamuna (YY)	Normal/control cross
6. Yamuna	X	Local (YL)	No heterosis exhibited
7. Local	X	Ganga (LG)	No heterosis exhibited
8. Local	X	Yamuna (LY)	No heterosis exhibited
9. Local	X	Local (LL)	Normal/control cross

No. of crosses (Diallele 2)

Female		Male	
1. Sutlej	X	Sutlej (SS)	Normal/control cross
2. Sutlej	X	Brhmaputra (SB)	No heterosis exhibited
3. Sutlej	X	Local (SL)	No heterosis exhibited
4. Brahmaputra	X	Sutlej (BS)	No heterosis exhibited
5. Brahmaputra	X	Brahmaputra (BB)	Normal/control cross
6. Brahmaputra	X	Local (BL)	No heterosis exhibited
7. Local	X	Sutlej (LS)	No heterosis exhibited
8. Local	X	Brahmaputra (LB)	No heterosis exhibited
9. Local	X	Local (LL)	Normal/control cross

VI. Back-cross and triple-cross hybrids

A number of back-cross and triple cross hybrids have been also produced and studied among the hybrids of Indian major carps by crossing the hybrid females with normal males of parent species. The kalbasu-mrigal female hybrid has been back-crossed with the normal males of *Labeo calbasu* and mrigal resulting in kalbasu-mrigal-kalbasu and kalbasu-mrigal-mrigal back crossed hybrids. Similarly kalbasu-rohu hybrid female has been back crossed with the normal male of *L. calbasu* producing kalbasu-rohu-kalbasu hybrid (Ibrahim, 1977). Again rohu-catla female hybrid has been back-crossed with the normal male of *L. rohita* resulting in rohu-catla-rohu hybrid.

Back-cross hybrids have been also produced by crossing the female hybrid, produced between catla female and male of rohu, with the male of catla (F1 back-cross). Again when specimens of the F1 back-crossed hybrid were crossed among themselves (F2 back-cross) five interestingly phenotypically different types of individuals were noticed (Sarangi Personal Communication). In the photographs of these hybrids, some were found to be intermediate to catla and rohu even in body color while some others appeared very close to rohu. The study is being continued (Sarangi pers.comm.).

Two triple-cross hybrids have been also produced by crossing the female hybrid of kalbasu-mrigal with the male of *C. catla* thus resulting in kalbasu-mrigal-catla and the female hybrid of kalbasu-rohu with the male of *Cyprinus carpio* giving rise to the kalbasu-rohu-*Cyprinus carpio*.

Almost all these crosses have shown normal percentage of fertilization and hatching. The spawn/fry of these hybrids can be reared to adult size.

The back-cross hybrid kalbasu-rohu-calbasu exhibited a high percentage of fertilization reaching over 98%. The hybrid possessed a ventral mouth with a fringed lower lip and two pairs of barbels. The color of the hybrid resembled that of kalbasu. The yearlings of these hybrids (adults) were characterized by a smaller head and deeper body. In overall appearance, this back-crossed hybrid showed a close resemblance to kalbasu (Chaudhuri, 1971).

List of Back-cross and triple-cross hybrids

FEMALE MALE	HYBRID	COMMENTS
1. Kalbasu-mrigal x *L. calbasu*	kalbasu-mrigal-kalbasu	Hybrid cross has shown good percentage of fertilization. The hybrids could be reared to adult size. No further informaton available.
2. Kalbasu-mrigal x *C. mrigala*	kalbasu-mrigal-mrigal	Hybrid cross has shown good percentage of fertilization. The hybrids could be reared to adult size. No further informaton available.
3. Kalbasu-rohu x *L.calbasu*	kalbasu-rohu-calbasu	Hybrid cross has shown higher fertilization (98%). Hybrids more resembled the kalbasu.
4. Rohu-catla x *L. rohita*	rohu-catla-rohu	Hybrid cross has shown good percentage of fertilization. The hybrid cross resulted in five different phenotypes with regard to body shap & colour..
5. Kalbasu-mrigal x *C.catla*	kalbasu-mrigal-catla	Fertilization in the hybrid was normal (70-80%). No other details available (needs further study)
6. Kalbasu-rohu x *C. carpio*	kalbasu-rohu-common carp	No details are available. (needs further study)

4. CYTOGENETIC STUDIES

The study of fish chromosome (karyotype) was initiated in India from 1960s by using basically the methodologies available for mammals. Gradually modified fish chromosome methodologies have been developed for obtaining quality metaphase chromosomes (Reddy and John 1986; Banerjee, 1987; Reddy and Tantia, 1992 and Nagpure and Barat, 1997). Of about 2,000 species of inland and marine fish analysed for karyological information (Das and Barat, 1995), over 200 species belong to India, which include both freshwater as well as marine species.

Most of these studies are related to the analysis of diploid karyotypes. Karyological studies on Indian major carps have been carried out by many workers (Khuda-Bukhsh and Manna, 1974; Majumdar and Ray Chaudhuri, 1976; Zhang and Reddy, 1991; Jana, 1993). However, regarding Indian major carps, the investigations, besides karyotype studies, also pertain to a comparative account of parental species and their hybrids which were attempted to interpret hybrid viability, fertility and sterility (Reddy *et al*, 1990a, Zhang and Reddy, 1991 and Jana, 1993).

4.1 Karyotype of Indian major carps

Karyotype studies of Indian major carps have been worked out by numerous scientists, who reported similar results. The type of chromosomes and the number of chromosomes in each type as reported by different workers are given in Table 7.

Table 7. Comparative karyological studies on Indian major carps as reported by various authors

Species	Type of chromosomes				Number of fundamental arms	Authors
	2n	Metacentric	Submetacentric	Subtelocentric		
L. rohu	50	18	8	24		Manna & Prasad (1971)
	50	6	26	18		Majumdar and Ray Chaudhuri (1976)
	50	10	16	24		Gui *et al.*, (1986)
		10	18	22	78	Zhang & Reddy (1991)
			18	22	88	Jana, 1993
C. catla	50	4	24	22 A/t		Khud-Bukhsh & Manna (1976)
	50	8	16	26		Manna (1977)
	50	6	32	12		Majumdar & Ray-Chaudhuri (1976)
	50	12	16	22	78	Zhang & Reddy (1991)
	50	12	16	22	88	Jana (1993)
C.mrigala	50	6	8	36 10st = 22A/t		Manna & Prasad (1971)
	50	6	26	18 (A/t)		Majumdar & Ray-Chaudhuri (1976)
	50	12	18	20 St/t	80	Zhang & Reddy (1991)
L.calbasu	50	6	8	36		Manna & Khuda-Bukhsh 1977
	50	6	32	12		Majumdar & Ray-Chaudhuri, 1976

4.1.1 *Catla catla* and *Labeo rohita*

The karyotypes of *Catla catla* and *Labeo rohita* have been studied by different workers (Khuda-Bukhsh and Manna, 1974; Mujumdar and Ray-Chaudhuri, 1976; Zhang and Reddy, 1991 and Jana, 1993). All these workers reported the diploid number (2n) in these two species as 50. In catla, the karyotype according to Zhang and Reddy (1991) consisted of 12 metacentric, 16 submetacentric, and 22 subtelocentric chromosomes. In rohu, the karyotype according to these authors consisted of 10 metacentric, 18 submetacentric and 22 sub-telocentric chromosomes. Later study with regards to comparative karyotype of these carps (Jana, 1993) also more-or-less agree with the observations of Zhang and Reddy (1991) (Table 7).

4.1.2 *Cirrhinus mrigala*

Karyotype studies on *Cirrhinus mrigala* have been performed by Manna and Prasad (1971); Majumdar and Ray-Chaudhuri (1976) and Zhang and Reddy (1991). All these studies have shown the diploid number is 50.

4.1.3 *Labeo calbasu*

The chromosomes of *L. calbasu* have been studied by Majumdar and Ray-Chaudhuri (1976), and Manna and Khuda-Bukhsh (1977a). The observations of these workers confirm the diploid number as 50 but differ with regard to the number of submetacentric and acentric types, (though they agree with the number of metacentric types).

4.2 Karyotype of hybrids

4.2.1 Indian major carp hybrids

Karyotype studies of hybrids within the species of Indian major carps have been studied only in the case of kalbasu-rohu, (Krishnaja and Rege, 1975); kalbasu-catla (Khuda-Bukhsh and Manna, 1976); rohu-catla (Khuda-Bukhsh and Manna, 1976; catla-rohu (Jana, 1993). In these hybrids the diploid number and type do not differ from the parent species, but the number under each type differ. Thus, the karyotype of the rohu-catla hybrids consists of four metacentric, 24 submetacentric and 22 acrocentric pairs which differs from the parent species, (Khuda-Bukhsh and Manna, 1976). The karyotype of the catla-rohu hybrid was reported to consist 12 metacentric, 10 submetacentric and 28 subtelo + telocentric (16+12 respectively) according to Jana (1993). In the hybrid between *L.calbasu* and *C.catla*, Khuda-Bukhsh and Manna (1976) recorded 10 metacentric, 14 submetacentric and 26 subtelocentric/ telocentric chromosomes. Krishnaja and Rege (1975) reported the diploid (2n) number of chromosomes as 50 in the case of the interspecific hybrid produced between *L.calbasu* and *L.rohita*.

4.2.2 Hybrids between Indian major carps and common carp

Regarding the karyological studies of these intergeneric hybrids, only the diploid number of chromosomes has been worked out and was reported to be = 76 in the case of *L.rohita* female and common carp male hybrid (John and Reddy, 1987). In the case of the hybrids between *C.carpio* female x *L. rohita* male; *C.carpio* female x *C.catla* male and *C.carpio* female x *Cirrhinus mrigala* male, the chromosome number varied from 74 to 76 in all these forms as mentioned earlier, which is around the sum total of the chromosomes of their parental gamets. However, karyotypes of about 75% of the sum total of chromosome numbers, in the case of *C.carpio* x *L.rohita*, 60% in *C.carpio* x *C.catla,* and 58% in *C.carpio* x *C.mrigala* hybrids showed 76, 74 and 75 chromosomes respectively. (Reddy *et al*, 1990). Details of karyotypes of these hybrids are in Table 8.

Khuda-Bukhsh *et al.* (1988) studied the karyotype of the suspected gynogen progeny resulting from the cross between *C.carpio* female x *L.calbasu* male. Karyotype of the progeny resulting from this cross was 2n=100 as in the case of the maternal parent, while 2n=50 in the case of the paternal parent *L.calbasu*. The morphology of the chromosomes in the F1 offspring (putative gynogen) was reported to be slightly different from that of the maternal parent *C.carpio*, though, their external anatomical features were similar. The haemoglobin and transferrin bands of F1 individuals and the parental forms suggest that the F1 individuals originated due to gynogenetic mode of reproduction by eliminating the genetic input from the sperm and diploidization of the ovum nucleus by the union of the nucleus from the polar body, thus limiting the role of the penetrating sperm to only activation. (Khuda-Bukhsh *et al.*, 1989). The appearance

have also exhibited the centromeric C-bands. The biarmed submetacentric chromosomes also revealed the presence of centromeric C-bands.

Attempts on G-banding of *L.rohita* did not meet with any interpretable success. The results were described as rather erratic and resolution of bands was not of good quality and also was not repeatable in subsequent attempts (Khud-Bukhsh and Tiwari, 1994).

Studies on NORs in *Catla catla* and in three species of the genus Labeo viz. *L.rohita, L.calbasu* and *L.bata* have been made by John *et al.* (1992 and 1993). In catla, the NORs of minute size were terminal to the short arms of one submetacentric and one subtelocentric chromosome pairs. In the case of *L.rohita, L.calbasu* and *L.bata*, the NORs were on the short arms of medium sized submetacentric chromosomes. The NORs in *L.rohita* and *L.calbasu* were similar in size, whereas those in *L.bata* were bigger in size. Again, while the NORs in *L.rohita* and *L. calbasu* were located on the 11th pair of chromosomes, in *L.bata*, they were on the 9th pair.

Chromosome banding studies are a step nearer to accuracy in enabling the identification of species and also homologous pairs in karyotypes. Similarly, banding of NORs also helps in this regard, as these are present on specific chromosomes, it makes relatively easy to distinguish species with similar karyotypes. (Sola *et al.*, 1984). The study also may help in identifying the parentage of putative hybrids that are sometimes found in nature.

The chromosome banding work particularly in fishes as a whole is very scanty. Of the 20,000 known species of fish in the world, banding studies appear to have been carried out in only 50 species. (Khud-Bukhsh and Tiwari, 1994). The reasons for this were many, of which the most important ones seem to be relatively smaller size and greater number of chromosome in fishes. Absence of proper and repeatable methodologies also appear to be one of the constraints in the progress of fish chromosome banding studies.

Regarding Indian carps,the main hurdle seems to be lack of adequate facilities for the workers especially in the Universities where such studies are generally carried out, to handle relatively bigger species like major carps. Indian major carps, unlike other small aquarium fishes, need farm (pond) facilities for handling, which are often not possible/available in the universities. As a result these researchers mostly concentrate on other commercially important catfishes or air-breathing fishes. Most of the fish cytological work, particularly in India is confined to such species of fish referred above as it is relatively easy to handle them. The advent of molecular level of research in genetics also appears to have diverted the attention and enthusiasm of many young and new researchers in genetics away from chromosome studies.

However, recently many of the research institutes are gradually taking up basic and academic research besides applied aspects. More and more laboratories are now concentrating on carp genetics research, especially pertaining to cytogenetics, biochemical and molecular genetics and also selection studies, not only in India but also in its neighbouring countries.

5. CHROMOSOME MANIPULATIONS

Through chromosome manipulations a new genetic status can be attributed to a given individual by altering or modifying its genome for better performance in its environment. These ehromosome manipulations or chromosomal engineering techniques involve tampering with the original genomic structure of a given individual in a systematic manner and abiding by certain biological laws (Reddy, 1992). Chromosome engineering may involve addition of extra set (s) to the existing complement, resulting in triploid or tetraploid individuals, or replacement with a duplicate set of chromosomes from the same individual, the resultant products being either gynogens with maternal inheritance or androgens with paternal inheritance.

5.1 Gynogenesis

Gynogenesis is the process of embryonic development with solely the maternal genome and without paternal genetic input, a phenomenon similar to parthenogenesis . Gynogenesis occurs in nature and can be also induced.

5.1.1 Natural gynogenesis

Natural gynogenesis has been reported in some members of the family Poecilidae, e.g. *Poecilia formosa* and *Cyprinidae* e.g. *Carassius auratus gibelio* (Gold, 1979). The cross between certain species of *Cyprinidae* such as the Chinese grass carp and silver carp also resulted in gynogenetic offsprings (Andriyasheva, 1973, Mantelman, 1973, Reddy, 1989 and 1995).

5.1.2 Artificial (induced) gynogenesis

Artificial induction of gynogenesis involves genetic inactivation or destruction of DNA content of milt by exposing it to either ultraviolet or gamma rays and activation of eggs with this genetically inactivated milt. However, these activated eggs develop into haploid embryos/hatchlings and do not survive unless diploidy is restored by subjecting the activated eggs to either thermal or pressure shocks at an appropriate stage after activation, just before the extrusion of the polar body (meiotic gynogenesis) or blocking the first mitotic division in the developing/activated egg (mitotic gynogenesis). The pace of development may differ from species to species and depends on the ambient temperature (Figure 6 & 7). Artificial gynogenesis has been successfully induced in many species of fish including carps (Golovinskaia, 1968; Purudom, 1969; Burlakov *et al.*, 1973; Stanley *et al.*, 1975; Chourrout D. 1980; John *et al.*,1984 and 1988; Wu *et al.*, 1986; Reddy *et al.*, 1993).

5.1.2.1 Induction of meiotic gynogenesis in Indian major carps

Meiotic gynogenesis was successfully induced in Indian major carps for the first time in *L. rohita* and *C. catla* in the year 1981 (John *et al*, 1984) and in *Cirrhinus mrigala* (John *et al.*, 1988, later by Hussain *et al.*, 1997). The procedure includes, genetic or DNA denaturalization or inactivation in the penetrating spermatozoa by exposure to either ultraviolet or gamma rays. For inactivating genetic component of the male gametes, ultraviolet irradiation is a simple device and very effective in fishes. To restore diploidy in the activated egg, either thermal stocks (heat or cold) or pressure shock are usually administered . However, the efficacy of these shocks, especially thermal shock, and their intensity and duration of treatment seems to vary from species to species. The effective intensity lies at some point around the sublethal levels. Thus in all the species of Indian carps, cold shocks below 10^{0}C and heat shocks above 42^{0}C proved to be lethal (John *et al.,* 1984 & 1988 and Reddy *et al.* 1993). Effective genetic inactivation of sperm DNA was possible in Indian major carps by exposing the milt, diluted to 1:4 with cold Hank's solution, (milt 1 and Hank's 4 parts), to a 15w ultraviolet germicidal tube at a distance of 20 cm and irradiated for 17-20 min (John *et al*, 1984). The milt was poured into petri dishes fixed on ice blocks, keeping the column depth of milt to 2mm and placing the petri dish on a mechanical shaker for irradiation. It is preferable to use milt from a different species since improper irradiation can be detected by the presence of hybrids. In the case of Indian major carps meiotic gynogenes of rohu were produced by activating the eggs with the irradiated milt of catla and vice versa (John *et al.*, 1984) and mrigal gynogens were produced by activating the eggs with the irradiated milt of common carp (John *et al.*, 1988).

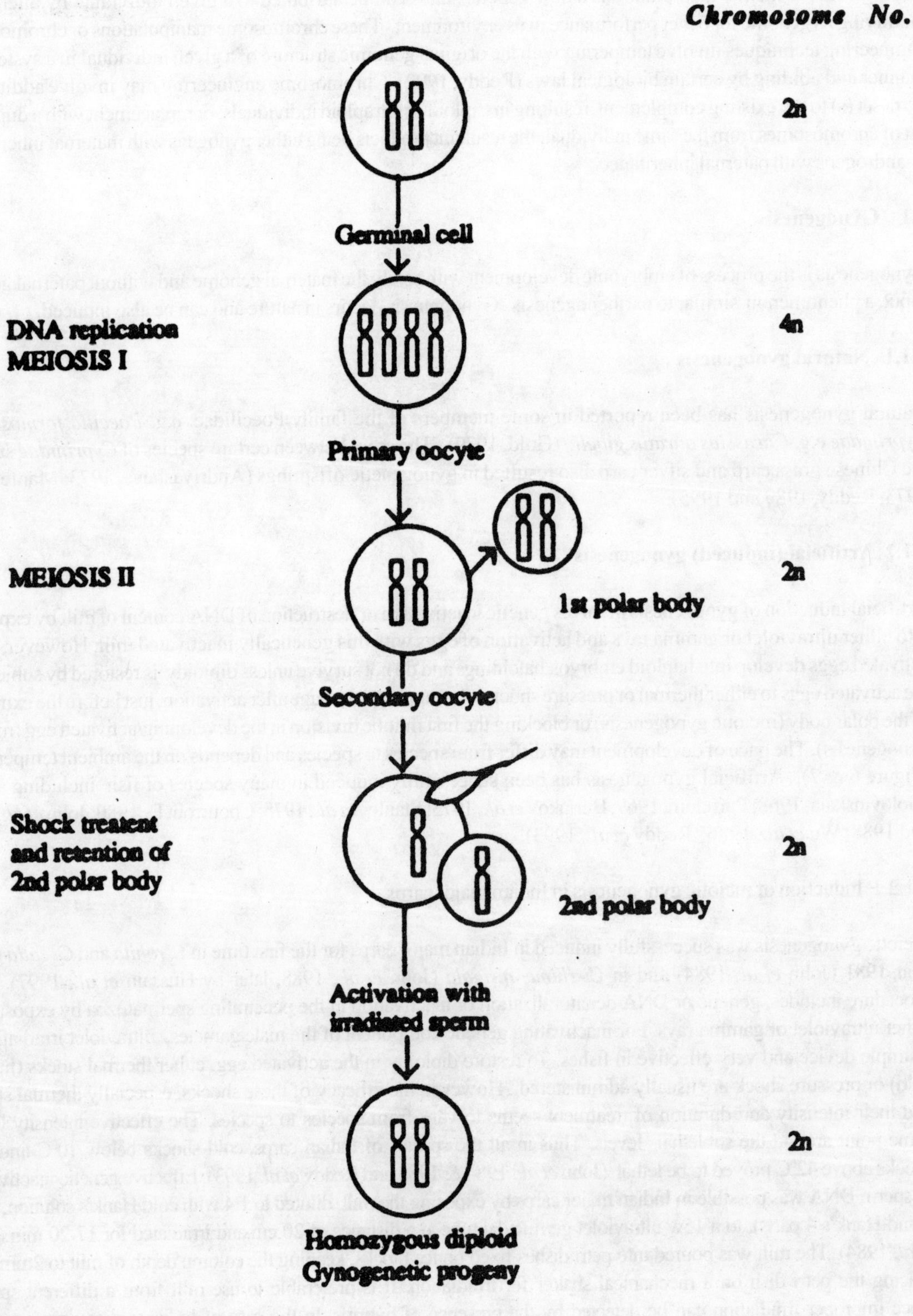

Figure 6. Illustration of events for inducing meiotic gynogenesis in Indian carps (Reddy *et al.*, 1997)

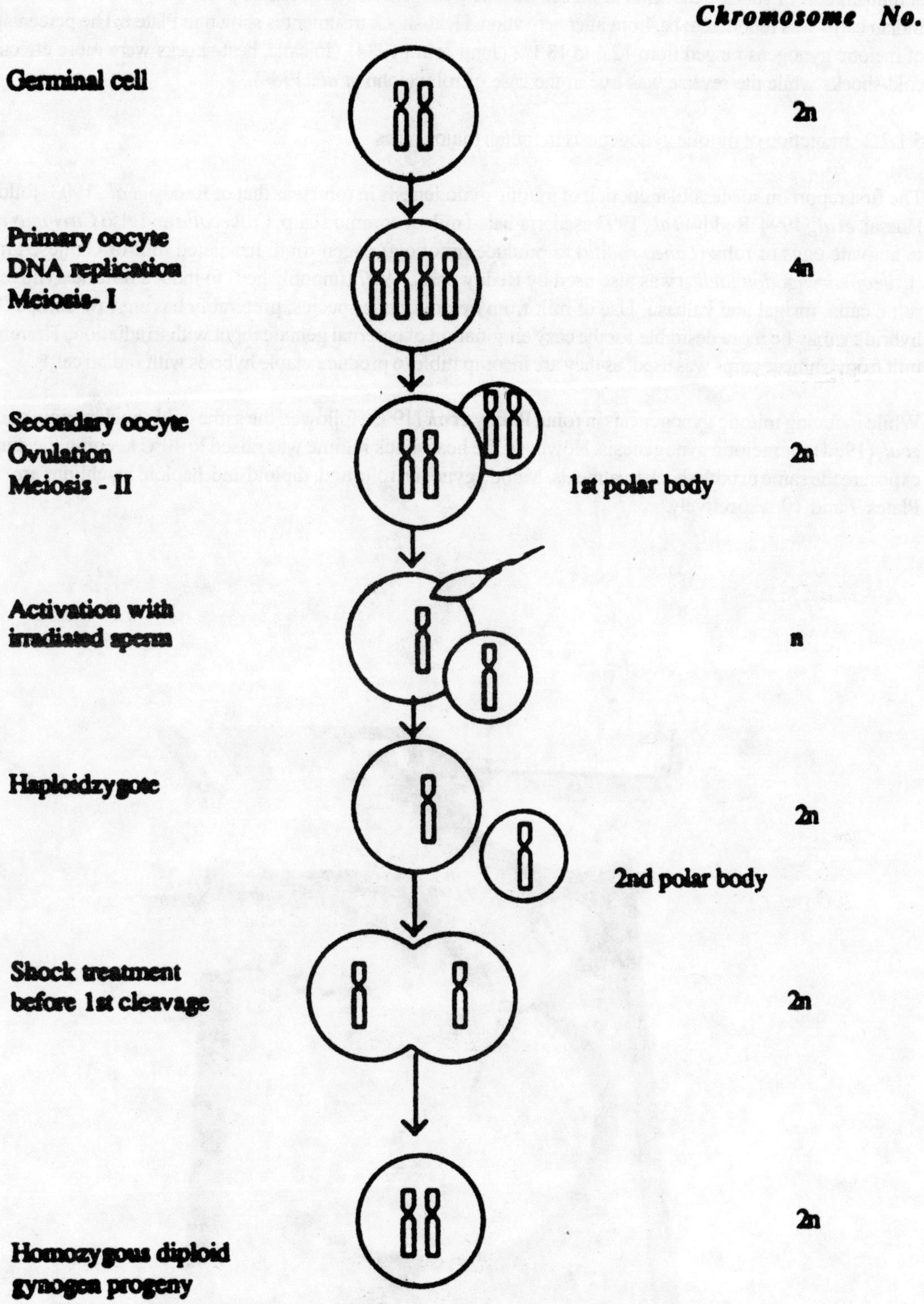

Figure 7. Illustration of events for inducing polyploidy (triploidy and tetraploidy) in Indian carps (Reddy *et al.*, 1997)

The effective intensities of cold and heat shocks for inducing diploid meiotic gynogenesis in Indian major carps were 12^{0}C for 10 min and 39^{0}C for one to two minutes respectively. The appropriate time after activation for the administration of shock treatments to induce meiotic gynogenesis (retention of polar body) in the activated egg of Indian carps was reported to be 4min after activation. Heat-shock treatment is shown in Plate 8.The percentage yield of meiotic gynogens ranged from 12.6 to 18.1% (John *et al.*, 1984). In catla, heat-shocks were more effective than cold-shocks while the reverse was true in the case of rohu (John *et al.*, 1984).

5.1.2.2 Induction of mitotic gynogenesis in Indian major carps

The first report on successful induction of mitotic gynogenesis in rohu was that of Reddy *et al.*, 1993, followed by Hussai, *et al.*, 1994. Reddy *et al.*, 1993 used irradiated milt of common carp, *Catla catla* and also *Cirrhinus mrigala*, to activate eggs of rohu (*Labeo rohita*) to produce mitotic gynogen rohu. Irradiated milt of Chinese grass carp (*Ctenopharyngodon idella*) was also used by Reddy *et al.*, 1997 (unpublished) to induce mitotic gynogenesis in rohu, catla, mrigal and kalbasu. Use of milt from heterologous species, preferably having low compatibility to hybridize may be more desirable for the easy elimination of paternal genetic input with irradiation. Therefore, the milt from Chinese carps was used, as they are incompatible to produce viable hybrids with Indian carps.

While inducing mitotic gynogenesis in rohu, Reddy *et al.* (1993) followed the same cold shock regime as did John *et al*, (1984) for meiotic gynogenesis. However, the heat shock regime was raised to 40^{0}C keeping the duration of exposure the same in both shock treatments. Mitotic gynogen rohu and, diploid and haploid hatchlings are shown in Plates 9 and 10 respectively.

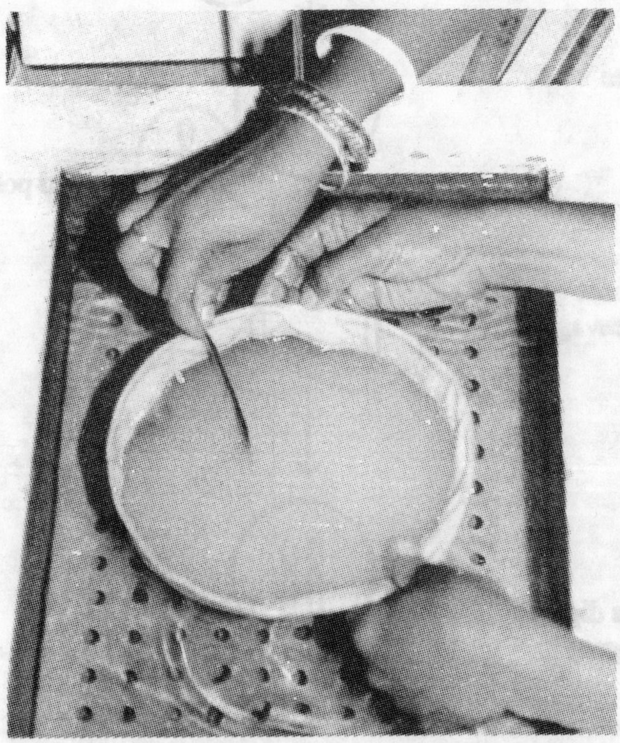

Plate 8. Heat-shock treatment to activated/fertilized eggs of Indian major carps to induce diploid gynogenesis/polyploidy respectively

Plate 9. Mitotic gynogen rohu *(L. rohita)*

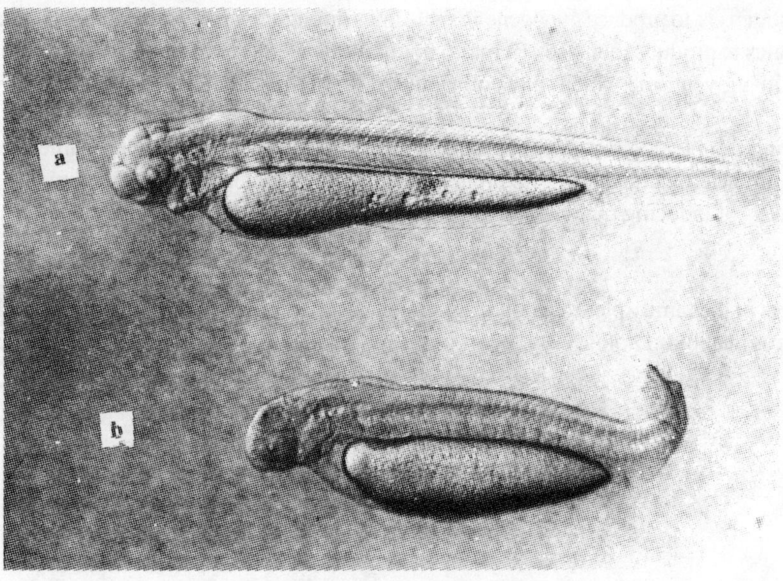

Plate 10. a. Diploid (normal (embryo)) b. haploid embryo (developed from an egg, activated by a genetically inactivated/denaturated sperm)

The yield of mitotic gynogens as reported by Reddy *et al.* (1993) was 15-30% up to fry stage and 0.66-10% up to fingerling stage with heat shocks and 30 and 4.3% up to fry and fingerling stages, respectively with cold shock treatments. The average survival rates ranged from 1.13% with heat shocks to 4.3% with cold shocks. Regarding the response of gynogens to treatment, John *et al.* (1984) and Reddy *et al.* (1993) showed that rohu responds better to cold shocks than heat-shocks, while inducing mitotic gynogenesis, as mentioned earlier, cold shocks yielded higher percentage of gynogens (4.3%) as compared to heat shocks (1.13%). However, the yield of meiotic gynogens in Indian carps was observed to be higher than that of mitotic gynogens as observed by John *et al.*, 1984 and Reddy *et al.*, 1993, respectively.

This trend appears similar in other species also. Purdom *et al.*, 1985 reported 1-7% survival rates in mitotic gynogen rainbow trout. The low yields in mitotic gynogenesis are attributed to the probable expression of deleterious or lethal genes in the highly homozygous state. Variations in the quality and performance of eggs from different females or from different batches of eggs of the same female have been also reported (Mair *et al*, 1986). However, the experience of different workers indicates that late shock treatments (mitotic gynogenesis) usually results in lower yields than those of early shocks (meiotic gynogenesis).

5.1.2.3 Meiotic versus mitotic gynogenesis

The qualitative difference between these two types of gynogenesis is that, it requires about 4-5 generations through meiotic gynogenesis to achieve complete homozygosity (Nagy *et al.*, 1984), whereas, the same result can be achieved with one generation by mitotic gynogenesis, as blocking of first mitotic division results in retention of two identical replicated mitotic products (Mair *et al.*, 1986).

Whether meiotic or mitotic, gynogenesis is an effective tool for producing inbred lines of fish in a much shorter time when compared to the conventional method of sib-mating.

The aim of developing inbred homozygous lines in Indian major carps especially *L. rohita* is to test the heterosis effect that may result by intraspecific hybridization. Nagy *et al.* (1984) reported heterosis effect in gynogen hybrids.

Another use of gynogens is to produce gynogenetic male lines through sex reversal of the gynogen females and use these functional males to produce mono sex (female) populations. Females of Indian carps grow larger than their male counterparts in a given time which may contribute to higher production.

The progeny resulting from gynogenesis will be usually all females where female homogamety exists. In the case where the females are heterogamous then the progeny should be 50% females and 50% males. A comparative account of the process of inducing meiotic and mitotic gynogenesis in Indian carps and the resultant products is given in Table 9.

Among the members of the family pleuronectidae, gynogenesis (haploid) occurs when plaice or flounder eggs are fertilized by milt from halibut. Diploidy can be induced by the application of cold shock to those haploid eggs (Purdom, 1983).

Table 9. Details of induced gynogenesis of carps at CIFA

Species	Gynogenesis		Shock treatment C.S. = cold shock H.S. = heat shock			Results
			Nature	Intensity	Duration	
C.catla	Meiotic	Mitotic	C.S.	12°C	10 min	Diploid gynogenesis
C.catla	Meiotic	Mitotic	H.S.	39°C	1 min	Diploid gynogenesis
L.rohita	Meiotic	Mitotic	C.S.	12°C	10 min	Diploid gynogenesis
L.rohita	Meiotic	Mitotic	H.S.	39°C	1 min	Diploid gynogenesis
C.mrigala	Meiotic	Mitotic	C.S.	12°C	10 min	Diploid gynogenesis
C.mrigala	Meiotic	Mitotic	H.S.	39°C	1 min	Diploid gynogenesis
L.calbasu	Meiotic	Mitotic	H.S.	40°C	1 min	Diploid gynogenesis
L.calbasu	Meiotic	Mitotic	C.S.	12°C	10 min	Diploid gynogenesis

N.B. Shock treatments were to 4 min ld inseminated eggs in all the cases.

5.2 Androgenesis

Androgenesis, results in progeny with only paternal inheritance. Like gynogenesis, androgenesis also occurs in nature and can be induced, though not as efficiently or easily as inducing the former.

5.2.1 Natural androgenesis

Natural or spontaneous androgenesis has been observed in certain hybrid crosses produced either between remotely related individuals or those with disproportionate or not so compatible genomes as in the cross between *Cyprinus carpio* female and grass carp male or *A. nobilis* x *Cyprinus carpio*, though, with a low incidence. Indian carps being highly compatible, among themselves, no such instances were ever reported in any of the hybrid crosses.

5.2.2 Artificial androgenesis

The process of inducing androgenesis involves the genetic inactivation of an egg and its fertilization with normal sperm of the corresponding species. The egg however, may develop into a haploid embryo and either thermal or pressure shocks have to be administered to restore diploidy as is done in the case of artificial diploid gynogenesis. The mechanism of restoring diploidy in androgenesis is quite different from that of gynogenesis. The process probably involves dispermy or other mechanisms.

Attempts of inducing androgenesis in Indian major carps have been made earlier, but were unsuccessful mainly due to the difficulties in the effective irradiation of the egg. The egg consists of a huge mass of cytoplasm and yolk that prevent the proper exposure of the chromosomes to UV rays in the nucleus which especially in carps, is oriented towards one side of the cytoplasm and usually facing downwards, covered by the cytoplasm and yolk. UV irradiation of eggs may need a longer duration which may be lethal. Irradiation under gamma rays may be effective in genetic denaturing of eggs in carps due to stronger intensity and lesser duration of exposure.

5.3 Polyploidy

An increase in the level of ploidy of an individual by the addition of one or more set(s) of chromosomes refers to polyploidy resulting usually either in triploidy or tetraploidy. Sometimes it may also result in penta or hexaploidy. Figure 8.

The phenomenon of polyploidy has been observed to take place in nature as in the case of gynogenesis and androgenesis. It can be also induced artificially.

5.3.1 Natural polyploidy (triploidy/tetraploidy)

Polyploidy has been observed to occur in nature in some species of fish like the common carp (*Cyprinus carpio*) and trout mainly due to chromosomal translocation and when two distantly related fish species are crossed. The crosses between grass carp and bighead carp had produced triploid hybrids (Marian and Krasznai, 1978). However, none of the interspecific nor intergeneric hybrid crosses among Indian carps have been reported to produce such allotriploids.

5.3.2 Artificial induction of polyploidy in Indian major carps

Some preliminary attempts have been made to induce artificial induction of triploidy and tetraploidy in Indian major carps with varied degrees of success (Reddy *et al.*, 1987; Reddy *et al.*, 1990).

Reddy *et al.*(1987), made the first attempt of inducing polyploidy in one of the Indian major carps, *Labeo rohita* by using colchicine. They could induce only tetraploid and diploid mosaics. However, Reddy *et al.*(1990a) successfully induced triploidy in rohu, *L.rohita*, and tetraploidy in *L.rohita* and *Catla catla* by using thermal shocks. Similarly successful induction of tetraploidy was also reported in the case of *Cirrhinus mrigala* and *L. rohita* (Zhang, 1990). Triploidy was induced in rohu by administering heat shocks to zygotes, seven minutes after fertilization at 42 C +0.5 C for a duration of 1-2 minutes. However, the incidence of triploidy was only 12% (Reddy *et al.*, 1990). Most of the eggs died probably due to lethal temperature level. These authors also reported that rohu zygotes exposed to heat shock, prior to first cleavage at 39C+0.5C for two minutes yielded 70% tetraploids and those exposed to cold shocks of 10-15C for 10 min. yielded only 30-55% tetraploids. In the case of catla, heat shocks to zygotes at 40C exposed for two minutes prior to first cleavage yielded 30-65% tetraploids, but cold shocks were not effective.

Zhang (1990) induced tetraploidy in *C.mrigala* for the first time and also in *L.rohita*. He reported that heat shocks administered at 39-40C for two minutes in embryos 22-25 min post-fertilization could induce tetraploidy in mrigal and rohu. However, the percentage of polyploids ranged from only 10-40. Islam *et al.* (1994) reported induction of triploidy in *L. rohita* by applying heat shock at 40C for 2 min starting 4 min after fertilization. The rate of triploidy induction was found to be 60% . The survival in the heat shock groups was recorded 15% and 25% in the control groups. The growth rate of triploid individuals was found to be significantly higher than in diploids for weight and length, $P<0.01$ and $P<0.05$ respectively. Details of induced triploidy/tetraploidy in major carps are given in Table 10.

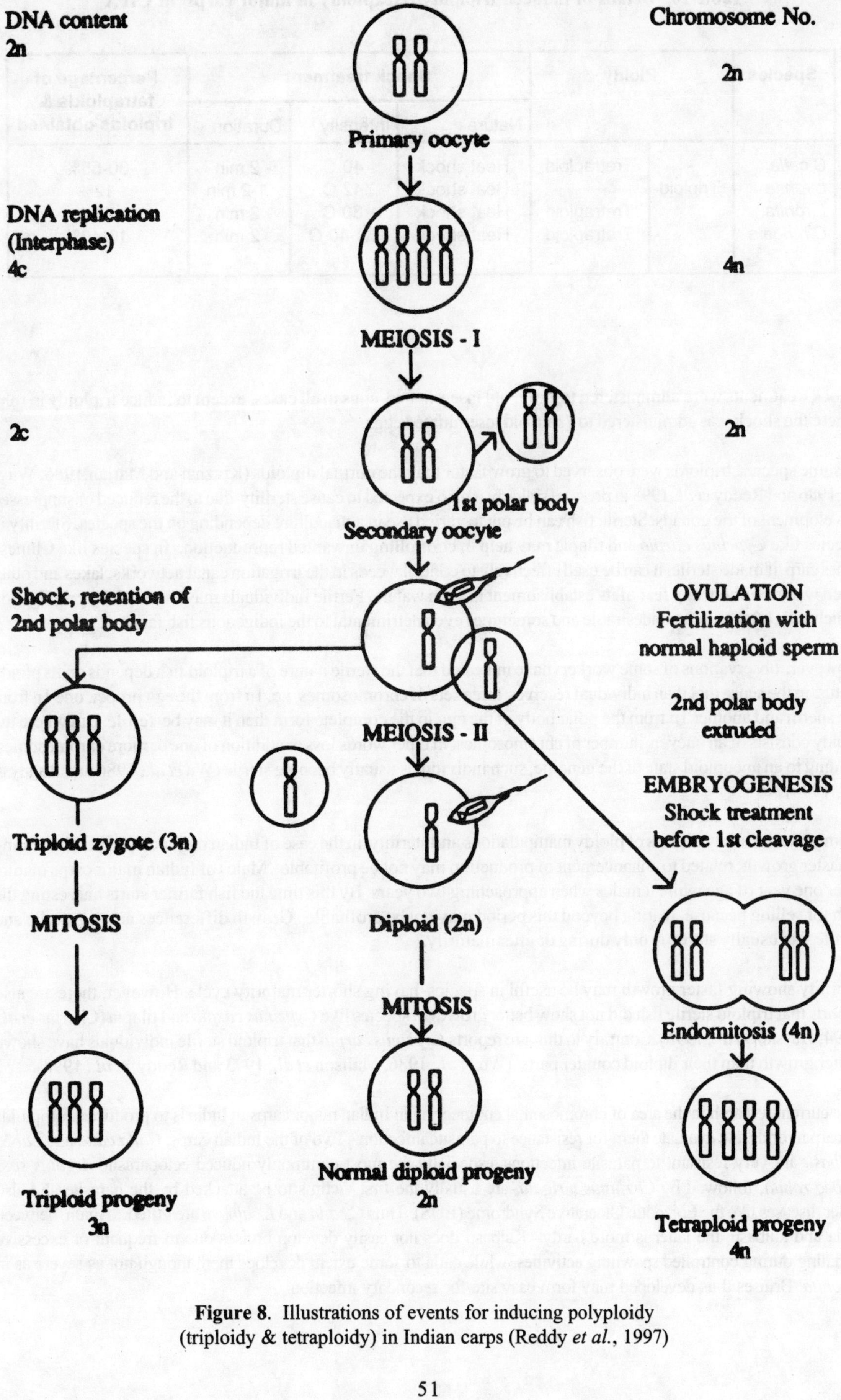

Figure 8. Illustrations of events for inducing polyploidy (triploidy & tetraploidy) in Indian carps (Reddy *et al.*, 1997)

Table 10. Details of induced triploidy/tetraploidy in major carps at CIFA

Species	Ploidy		Shock treatment			Percentage of tetraploids & triploids obtained
			Nature	Intensity	Duration	
C.catla	-	Tretraploid	Heat shock	$40^{\circ}C$	2 min	30-55%
L.rohita	Triploid		Heat shock	$42^{\circ}C$	1-2 min	12%
L.rohita		Tretraploid	Heat shock	$30^{\circ}C$	2 min	70%
C.mrigala		Tretraploid	Heat shock	$39\text{-}40^{\circ}C$	2 min	10-40%

Shock treatments were administered to 4 min old inseminated eggs in all cases, except to induce triploidy in rohu where the shock was administered to 7 min old inseminated eggs.

In some species, triploids were observed to grow faster than the normal diploids (Kraznai and Marian,1986; Wu *et al.*,1986 and Reddy *et al.*,1998 in press). Triploidy is also expected to cause sterility due to the reduced or suppressed development of the gonads. Sterile fish can be put to varied use in aquaculture depending on the species. Sterility in species like *Cyprinus carpio* and tilapia may help in controlling unwanted reproduction. In species like Chinese grass carp if made sterile, it can be used effectively to control weeds in the irrigation canal networks, lakes and other open waters without any fear of its establishment in such waters. Fertile individuals may establish by propagation which very often may be undesirable and sometimes even detrimental to the indigenous fish fauna.

However, observations of some workers have indicated that the sterile nature of a triploid fish depends on its ploidy status. in the sense that if an individual receives three sets of chromosomes, i.e. 1n from the egg proper, one 1n from the sperm and another 1n from the polar body of the egg in the complete form then it may be fertile. But where the ploidy consists of an uneven, number of chromosomes, in other words loss or addition of one or more chromosomes, leading to an aneuploid state of the genome, such individuals usually become sterile (Wu *et al.*, 1986 and Reddy *et al.*, 1990).

Coming to the utility aspects of ploidy manipulations and sterility, in the case of Indian major carps, sterility aiming at faster growth, related to enhancement of production may not be profitable. Males of Indian major carps mature after one year of age while females when approaching two years. By this time the fish farmer starts harvesting the fish for selling because, rearing beyond this period may not be profitable. Growth differences among sterile and fertile fish usually show up only during or after maturity.

Sterility showing faster growth may be useful in species having shorter maturity cycle. However, there are also reports that triploid sterile fish did not show better growth in species like *Cyprinus carpio* and tilapia (Cherfas *et al.*, 1994; Hussain *et al.*, 1995). Contrary to this are reports *Cyprinus carpio* that triploid sterile individuals have shown better growth than their diploid counter parts. (Wu *et al.*, 1986, Malison *et al.*, 1993 and Reddy *et al.*, 1998).

The current research in the area of chromosomal engineering in Indian major carps in India is to produce allotriploids of carp hybrids and evaluate them for resistance to parasite infection. Two of the Indian carps, *Catla catla* and *Labeo calbasu* are very resistant to parasite infections, especially the most commonly noticed ectoparasite *Argulus sp.* . *Labeo rohita*, followed by *Cirrhinus mrigala*, are usually the first victims to be attacked by the parasite and also other diseases like the Epizootic Ulcerative Syndrome (EUS). Thus *C.catla* and *L.calbasu* are gifted. Again, between catla and kalbasu, the latter is more hardy. Kalbasu does not easily develop bruises due to frequent or excessive handling during controlled spawning activities, while catla to some extent develops them though not as severe as in *L.rohita*. Bruises thus developed may form easy site for secondary infection.

Because of the above mentioned traits of catla and kalbasu, it is planned to produce allotriploids with these carp species by developing hybrids with males of the weaker species like the rohu and increasing the genome input from the maternal species with better resistance and then study the nature of the allotriploids. Moreover, Mishra (1991) reported that the rohu-catla and the reciprocal hybrids are generally prone to parasite infection, unlike catla. This may be due to the reduced genomic content of catla in the hybrid.

Since the hybrids of rohu and catla are reported to have better traits (Chaudhuri, 1973, Bhowmick *et al.*, 1981 and Konda Reddy and Verghese 1980 a, 1980 b), it is worthwhile to develop a resistant variety of these hybrids through allotriploidization. The hybrids between *L. rohita* and *L. calbasu* may also be similar. However, the incidence of parasitic infection is relatively less in these hybrids when compared to the former hybrids. i.e. the rohu-catla and the reciprocal hybrids.

5.4 Gene transfer or genetic engineering in Indian major carps

Gene transfer or genetic engineering in Indian major carps was first initiated by introducing growth hormone gene in the embryos of *L. rohita* (Alok *et al.,* 1995). This is the earliest work initiated on Indian major carp. These workers microinjected about 150,000 to 200,000 copies of a fusion gene containing marine metallothionine promoter (Mu MT) and the coding sequence of human growth hormone (hGH) into the fertilized cone of the embryos of *L. rohita.*

In total, about 434 embryos were microinjected, of which only 50 embryos survived. Ultimately, only 5 of these 50 embryos survived up to 90 days. Analysis of DNA from gill tissues of these surviving embryos has shown that only two of them contained transgene in their genome, carrying 1-2 copies of the human growth hormone gene in their genome. However, these results indicate that it is possible to introduce foreign DNA into the embryos of *L. rohita* (Alok *et al.*, 1995).

Reddy *et al.*, 1991 worked out some techniques to prolong the developmental events (cleavage) during the embryonic developments in major carps in order to facilitate gene transfer. Depending on the ambient temperature (28-30^{0}C), one celled stage (formation of blastodisc) appears in the fertilized eggs of major carps, 20-25 min after fertilization/ insemination and lasts only for 8-10 min. after which the first cleavage commences (2-celled stage).

Reddy *et al.* (1991) developed techniques for facilitating the implantation of foreign gene (DNA) directly into fertilized carp egg through microinjection. The usual time of development for the formation of one celled stage in the fertilized eggs could be prolonged for 20-30 min by incubating the eggs in water 20C rather than the normal tap or pond water at 29-30C.

Secondly, by keeping the female and male gametes separately in cold Hank's solution with proper dilution (gametes to Hank's solution 1:4, John *et al.*, 1984), it was possible to maintain the gametes viable for 2-3 hrs under refrigeration at 7-8^0 C or lower temperature than normal. This method allows researchers to freshly fertilize the eggs in batches at half an hour intervals, so that when one batch is injected, another batch can be made ready. Yet another method suggested was to inject female brood fish at different time intervals, so that when the fertilized eggs of one fish are injected, another fish will be ready to ovulate (Reddy *et al.*, 1991).

At present some more laboratories in India, such as the Centre for Cellular and Molecular Biology (CCMB), and Madurai Kamaraj University are engaged in genetic engineering studies, particularly in fishes including major carps. Isolation of growth hormone genes in *L. rohita* and *C. catla* has been also successfully done (Majumdar, 1997, Personal communication).

6. SELECTIVE BREEDING

Selective breeding proved to be the most effective process for genetic improvement of plants and animals. Achievements in increased yields of paddy, wheat, milk, meat etc. during recent decades are due in large to selection. In aquaculture, the potential has not been fully exploited through selection. However, the selection work on Atlantic salmon in Norway, the Channel catfish in the USA and the Nile tilapia in the Philippines is enough testimony to advocate that it is quite possible to improve many economic traits like, survival, growth, resistance to disease, meat quality etc. through selection. In fact, many hatchery breeding programs result in lower quality fish through inbreeding depression.

In most of the developing countries, particularly in India and other South-east Asian countries, a substantial number of people depend on fishing or fish culture for their living. Consequently, there is over-exploitation due to heavy fishing both in marine and riverine areas reducing the size of populations in the absence of proper replenishment of the depleted stocks.

In aquaculture ever since the success of induced breeding of the Indian major carps through hypophyzation, the technique has been gradually popularized throughout the Indian subcontinent as well as neighbouring countries such as Pakistan, Bangladesh, Burma, Nepal etc. Consequently, many hatcheries have developed in almost all these countries. However, in over four decades, attention has been paid only to produce fish seed to meet a pre-targeted quantity with very little regard to quality (Eknath 1991).

A survey of carp hatcheries in the state of Karnatak in South India was made (Eknath and Doyle 1990) because of the growing concern that hatchery stocks may be inbred, as suggested by rapid deterioration of their performance. Eknath and Doyle (1990) found that the rate of the accumulation of inbreeding (F) was particularly high for the three most desirable carp species in the country, catla, rohu and mrigal. The most important reasons for this appeared to be that all the hatcheries function as isolated genetically closed units, raising their own stocks of brood fish. Moreover, the individuals for replenishment of brood stock in the hatchery are those which were retained after the hatchery had reached its seed production target for a given year (Eknath and Doyle 1990). It was also reported that small hatcheries appear to rapidly inbreed their stocks. The tendency to breed slow growing and later maturing individuals was also observed in some of the hatcheries (Eknath and Doyle, 1985), which is undersirable.

The survey Basavaraju *et al.* (1997) conducted particularly for catla in the three major seed producing hatcheries in the state of Karnataka, namely Tunga Bhadra Dam (TBD), Bhadra Reservoir Project (BRP) and Kabiri Reservoir Project (KRP), also indicated that the practices followed by these hatcheries are likely to result in both inbreeding and negative selection.

Maheshwari and Biradar (1998) reported decline in fertility of hatchery bred *Labeo rohita* due to inbreeding. A study was conducted at Powarkheda, Freshwater Fish Farm of the Central Institute of Fisheries Education (ICAR), India where induced breeding of *L. rohita* was carried out between 1982 and 1992 to assess fertility. It has been reported that breeding of related individuals up to the sixth generation led to 71% decline in fertility level as compared to base population. It was also observed that the range, standard deviation and coefficient of variation were highest for fecundity in the first generation. This indicates, according to the authors, the scope for improvement through selection.

6.1 Selection work on Indian major carps in India

Currently two projects are in operation on carp selection. One on *Labeo rohita* as already mentioned, the other on *Catla catla* at the University of Agricultural Sciences, Bangalore, Karnataka State.

6.1.1 Selective breeding of rohu

Founder populations of *L. rohita* have been collected from five different rivers in North India, namely, Rivers Ganga, Gomati, Yamuna, Sutlej and Brahmaputra, which are the native habitat of rohu. Rohu stock from the fish farm of CIFA has also been added to the gene pool. Thus the base population for the selection project consisted of rohu, from six sources, to broaden the gene pool. Different stocks of rohu are shown in Plate 11.

Plate 11. Rohu *(L. rohita)* from different source

a. River Ganga *b. River Gomati*
c. River Yamuna *d. River Sutlej*
e. River Brahmaputra *f. Local (Farm) rohu*

All these stocks were collected from fry to fingerling stages and reared separately in earthen ponds for a period of one to two months before they were physically marked and stocked in communal ponds for raising to brood fish.

Marking was done using dye, fin clipping and a combination of both. Ganga rohu were dye injected with M-procian blue at the base of the right pectoral fin and Yamuna rohu at the base of the left pectoral fin. In the case of Gomati and farm rohu the right and left pelvic fins were clipped respectively, while the Sutlej rohu were marked with the blue dye, inject at the base of the left pectoral fin by clipping the right pelvic fin. The reverse has been done to mark the Brahmaputra rohu, i.e. by injecting blue at the base of the right pectoral and clipping the left pelvic fin (Gupta , 1994). During 1993, the first groups of full sibs were produced from the local and Ganga stocks as these two stocks were matured. Two or three males were mated with one female (paternal half sibs). However, during the year 1994, full-sib groups (families) were produced from all the six stocks of rohu: two or three females were mated to a male, producing maternal half-sibs. (Reddy *et al.*, 1998). Marking of rohu stocks has been shown in Figure 9.

Rearing full-sib groups to tagable size fingerlings has been done in earthen nursery ponds of 100 m^2 for four to five months (Saha *et al.*, 1997). After the full-sib attained suitable tagging size (8-10 g) the fish were tagged by using Passive Integrated Transponder (PIT) tags, before they are stocked in ponds for communal rearing under both mono and polyculture systems. (Mahapatra *et al.*, 1998). PIT tags are shown in Plate 12.

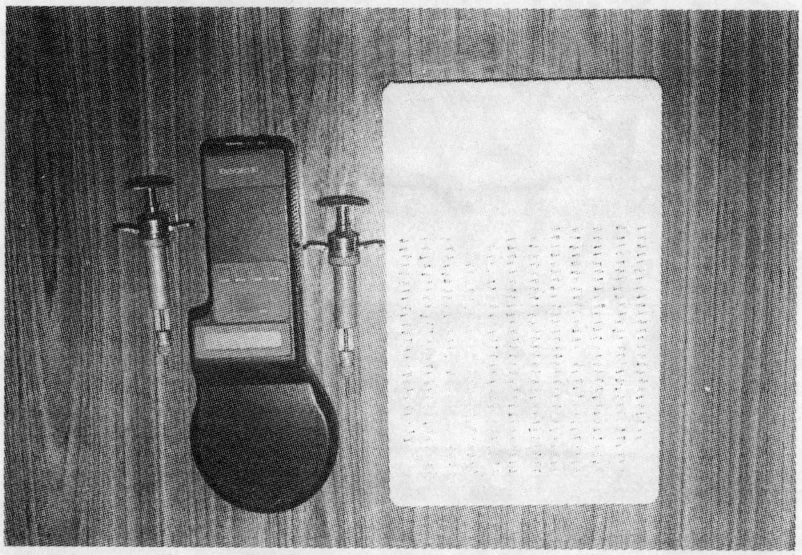

Plate 12. Passive integrated transponder (PIT) tags, tag implanter and tag reader

After a period of one year, data pertaining to growth (length and weight) was collected and analysed to evaluate the individual breeding value and to rank the fish. The data so far collected indicated that growth and survival of the wild rohu stocks were equal to or better than the farmed stock. This indicates that the present procedures followed by hatcheries for seed production are improper. A substantial additive genetic variation has been observed for growth rate and survival of rohu, both under mono and polyculture systems, thus indicating that growth rate and survival of rohu can be improved through selective breeding. Competition for food may prohibit rohu from exhibiting its growth potential in polyculture. Proper feeding and pond management is essential while testing fish in a breeding programme.

The study so far has also indicated that stocks and full-sib groups within stocks ranked consistently high for growth performance in mono and polyculture, which suggests that there is no need for the development of special rohu lines for different systems. However, further studies are envisaged to test genotype by environment interaction, i.e. under intensive/extensive production systems and in different climatic zones (CIFA and AKVAFORSK, 1996).

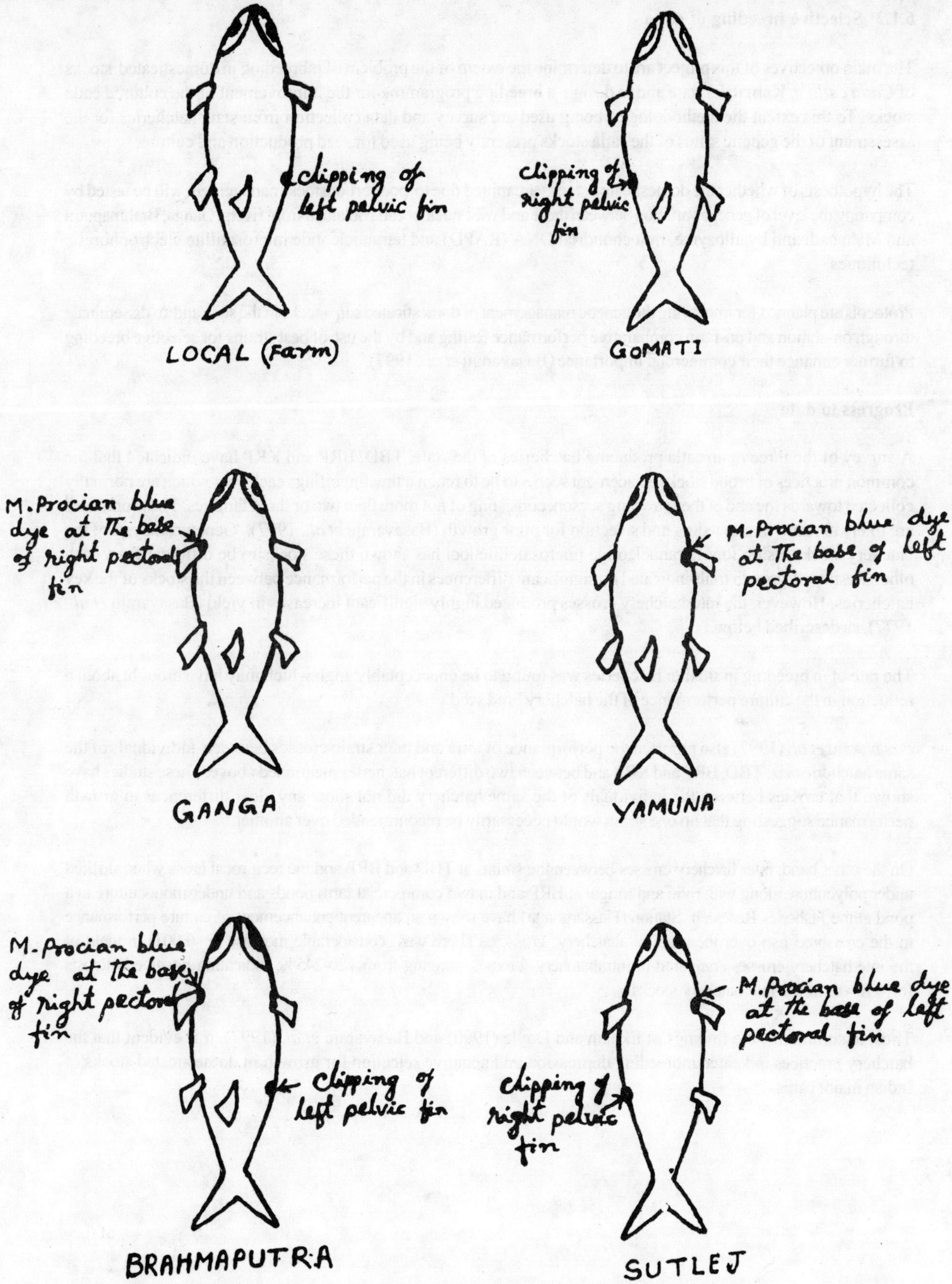

Figure 9. Illustration of marking rohu from different sources, with dye and fin clipping

6.1.2 Selective breeding of catla

The main objectives of this project are to determine the extent of the problem of inbreeding in domesticated stocks of *Catla catla* in Karnataka state and to design a breeding programme for the improvement of the cultured catla stocks. To this extent the methodologies being used are survey and data collection from state hatcheries for the assessment of the genetic status of the catla stocks presently being used for seed production and culture.

The hypothesis of whether the domesticated stocks are inbred due to poor broodstock management will be tested by comparing the level of genetic variation between these and wild stock of catla obtained from rivers, Ganga, Brahmaputra and Mahanadi and by allozyme, mitochondrial DNA (RAPD) and tetranucleotide microsatillite electrophoretic techniques.

Protocols are planned for improving the genetic management of domesticated carp stocks in the state and to disseminate through on-station and on-farm comparative performance testing and by the use of best strains for selective breeding to further enhance their commercial importance (Basavaraju *et al*., 1997).

<u>Progress to date</u>

A survey of the three main catla producing hatcheries of the state, TBD, BRP and KRP have indicated that the common practices of brood stock replacement seems to be to retain a few fingerlings each year which are normally collected towards the end of the spawning season, consisting of not more than two or three families. These practices are likely to result in inbreeding and selection for poor growth (Basavaraju *et al*., 1997). Genetic analysis of the hatchery stocks using four tetranucleotide microsatellite loci has shown these stocks to be different from each other. Results of growth trials indicated no significant differences in the performance between the stocks of the key hatcheries. However, the interhatchery crosses produced highly significant increases in yield (Basavaraju *et al*., 1977), as described below.

The rate of in breeding in smaller hatcheries was found to be unacceptably high which may have brought about a reduction in the culture performance of the hatchery bred seed.

Basavaraju *et al*. (1997) also reported the performance of intra and inter strain crosses between individuals of the same hatcheries viz. TBD, BRP and KRP and between two different hatcheries mentioned above. These studies have shown that crosses between the individuals of the same hatchery did not show any clear differences in growth performance suggesting that no one strain would necessarily be recommended over another.

On the other hand, inter hatchery crosses between the strains at TBD and BRP and the reciprocal cross when stocked under polyculture along with rohu and mrigal at BRP and in two commercial farm ponds and under monoculture in a pond at the Fisheries Research Station (Hassarghatta) have shown an apparent enhancement of culture performance in the crossbred fish over the pure intrahatchery crosses. There was considerable increase ($P<0.001$) in yield of the interhatchery crosses compared to intrahatchery crosses ranging from 120-245%, indicating positive heterosis for growth in the interhatchery stocks.

Thus, to summarize the findings of Eknath and Doyle (1990) and Basavaraju *et al*. (1997), it is evident that the hatchery practices indicated inbreeding depression and negative selection for growth in domesticated stocks of Indian major carps.

7. STATUS OF GENETIC RESEARCH ON INDIAN MAJOR CARPS IN OTHER COUNTRIES

The most actively involved countries, besides India, on genetic research of Indian major carps are Bangladesh, Vietnam and Thailand. Information is generally lacking regarding genetic work in Pakistan and Burma.

7.1 Bangladesh

Research on genome manipulation is being carried out and successful induction of meiotic and mitotic gynogenesis was achieved at the Fisheries Research Institute Freshwater Station, Mymensingh. Clonal lines of *L. rohita* produced by meiotic gynogenesis in the F_2 generation were reported to have been produced. (Hussain *et al.*,1994 b and Hussain *et al.*, 1997).Successful induction of triploidy was also reported in *L. rohita* (Islam *et al.*, 1994).

Bangladesh has also initiated selective breeding work on *C. catla* for genetic improvement. For this, catla were collected from different river systems and genetic characterization is being carried out by applying electrophoresis and morphological methods.

As indicated by the surveys conducted in Indian hatcheries (Eknath and 1990 and Basavaraju *et al.*, 1997), the investigations in Bangladesh have more or less shown the same trend. The existence of 400 carp seed producing hatcheries has been reported from Bangladesh. These hatcheries have been engaged in the production of seed since the 1970s. Some 31 billion spawn are reportedly produced annually. Problems mentioned with regard to hatchery produced seed in India have been also common in Bangladesh. Deterioration of hatchery bred stocks as indicated by retarded growth, morphological deformities and increased incidence of disease has been reported in recent years due to poor brood stock management. The data collected from several fish hatcheries located in Jessore, Comilla and Mymensingh revealed that most of the hatchery operators have no basic idea about brood stock management such as recruitment of new breeders at periodic intervals in the stock, optimum density of the brood stock and their proper feeding (Hussain and Mazid, 1997). These hatchery procedures lead to unconscious negative selection due to the use of undesirable size of brood fish and generation after generation of mating the related spawners from a small population.

Currently, Bangladesh is engaged in characterization of different land races of catla and rohu by the assessment of meristic and morphological characters, genetic fingerprinting of riverine stocks of catla using molecular techniques and karyotyping to assess ploidy of rohu and mrigal. Developing breeding plans for stock improvement of catla and induction of gynogenesis and production of genetic clones in rohu are part of a genetic improvement programme for carps in Bangladesh.

7.2 Thailand

Thailand has an elaborate genetic improvement programme, not only on carps but also for other species such as tilapias, catfishes, freshwater prawns, seabass, oyster, and aquarium fish. Among Indian major carps *L. rohita* has been currently taken up in Thailand for genetic work. Nineteen pairs in 1968 and 250 fish in 1969 were introduced into the country. The species is propagated through hatchery breeding and stocking in reservoirs and swamps where they have been already become established. It is also cultured in ponds. The production of rohu in 1994 was estimated at 1,783 tones ranking third in both weight and value among the carp species cultured in Thailand (Nuanmanee Pangthana, 1997) .

Genetic studies on rohu have been carried out with regard to the growth rate of the species raised in cages and to estimate the heritability as well as the genetic, phenotypic and environmental correlation for body weight, total length, body depth, head length and girth of fish at different ages (Dupsuksamran, 1987).

Mrigal was introduced into Thailand during 1980 and 82. A consignment of 100 fish was first introduced from Bangladesh. Later in August 1982 eight pairs of brood fish weighing 700-800 g and 500 fingerlings were brought to Kalasin Inland Fisheries Station. Like rohu the propagation of mrigal is also carried out through hatchery breeding and the seed is stocked in reservoirs and swamps. The species has now become established and is available throughout the country.

Thailand wants to set up a National Breeding Programme for hatchery produced carps and tilapias; among the Indian carps, *L. rohita* has been selected. The Central Institute of Freshwater Aquaculture (CIFA) is providing the improved rohu strain under the International Network for Genetics in Aquaculture (INGA) programme.

7.3 Vietnam

Vietnam, like India, Thailand and Bangladesh, is one of the member countries of INGA but also the carp genetic improvement programme in Asia. The Indian carp species included in carp culture practices in Vietnam are the rohu and mrigal.

Vietnam has an elaborate selective breeding programme for mrigal. Mrigal production is mainly from north Vietnam, but this species is not considered as important in the south. Of the 200, 000 tons production from freshwater, mrigal contribution was reported to be 45, 000 tons in north Vietnam (Bentsen *et al.,* 1996).

In north Vietnam, sexual maturity in mrigal occurs at two years of age as in India. The breeding process is repeated three times in a season. It has been reported that the first and the second breeding yield better quality seed than the third. This is because young and small brooders normally mature and spawn earlier in the season and, like in other regions, early produced fry fetch higher price. This tendency to prefer to use early mating and small sized brood stock may lead to selection for small size and early maturity.

The base population of mrigal consisted of about 1000 fingerlings, imported from only one source, i.e. the Thangon state fish Farm in Laos in 1984 to the Thu Due Fish Farm of Research Institute for Aquaculture No. 2 (RIA 2) with the support of the Interim Mekong Committee. Another consignment of 10 pairs of breeders were imported from the same source in 1986. At the Fish Farm of the Research Institute for Aquaculture No. 1 (RIA-1) about 4-5 generations have been produced from only 6 males and 6 females which were transported from RIA-2 from the original stock. Because of the repeated production from a limited stock of 6/6 brood fish, the growth rate was observed to decline year by year. Deformities have been consistently observed in about 5-7 percent of the fish. The fish also have been reported to mature at lower age and lower body weight.

Vietnam is planning to take up individual (mass) selective breeding programme for mrigal. However, before the start of the programme, the base population has to be upgraded. Thus, Vietnam has planned to directly import wild populations of mrigal from three different river systems in India, consisting of about 2000 fry from each river. The two main goals of Vietnam for aquaculture of mrigal carp are a high growth rate and a high proportion of females, as males are 60-70 percent of the female body weight at market. This is true in most other carps as well A detailed plan of the selection programme for mrigal in Vietnam has been drawn out by Bentsen *et al.* (1996).

8. CRYOGENIC PRESERVATION OF CARP MILT AND GENE BANKING

Protocols have been developed for sperm cryopreservation of major carps more or less simultaneously at the National Bureau of Fish Genetic Resource (NBFGR) and The Central Institute of Freshwater Aquaculture (CIFA) in 1988. The optimum requirements have been developed in respect to extender, activation media, dilution rate, activation period, sperm and egg ratio difference between species. Stocks of *L.rohita* milt cryopreserved in 1988 are being maintained along with other species in a mini gene bank at NBFGR. Viable hatchlings have been produced from five-year old rohu cryopreserved milt (Ponniah, 1997).

The ionic composition of seminal and variation fluid of rohu and mrigal have been studied at the Bureau with the objective of providing basic information that can be used to fashion new extenders and spermatozoa motility inducing media. Gupta and Rath (1993) preserved the milt of *C. catla, L. rohita* and *Cirrhinus mrigala* under cryogenic condition for three years. The quality of milt samples were analysed when preserved and were expressed in terms of spermatocrit value and sperm cell count, which in catla, rohu and mrigal ranged from 65-75, 75-85 and 65-75 respectively while sperm cell counts ranged between 2-2.5 X 10^7, 3.0-3.25 X 10^7 and 2.0-2.5 X 10^7 respectively. The most effective diluent was found to be an electrolyte medium when Dimethyl Sulfoxide (DMSO) acted as cryoprotectant and the ratio of milt and diluent was 1:4. Quick Thawing procedure (38 + 2^{0}C).

Gupta and Rath (1995) have also studied the efficacy of cryopreserved milt in fertilizing the eggs compared with normal milt. The survival and growth rates from larvae to fry and to fingerling produced from eggs fertilized with cryopreserved and fresh milt were not found statistically different.

The fertilization efficiency of cryopreserved spermatozoa thawed at different temperatures of catla, rohu and mrigal over a period of four years was assessed at different time intervals after thawing, (Gupta *et al*., 1995). Milt thawed at 39 + 1C and 30 + 1C had fertilization rates of 56.7% and 12.5% during 900 and 1500 seconds. The motility and fertility were reduced drastically at the next 60 seconds and 300 seconds.

8.1 Cryopreservation of embryos

Cryopreservation of embryos of *Cyprinus carpio* and *L. rohita* have been attempted (John *et al*., 1993). It has been reported that *C. carpio* at tail bud stage was more tolerant to cryoprotectants than morula stage as revealed by the hatching rates of treated embryos. This may be due to some ontogenic differences in the tolerance to handling, membrane permeability or osmoregulation potential, rather than cryoprotectant susceptibility. Rohu tail- bud embryos treated in methanol and DMSO (Cryoprotectents) in different concentrations and in combination, showed high hatching rates in all the treatments but no pattern was evident. When embryos from all treatments were kept immersed in liquid nitrogen for about 3 hours and thawed, no survival was observed. This was felt due to defective thawing (John *et al*., 1993).

Cryoprotectant toxicities have been also worked out in two embryonic stages of prawns viz. *Penaeus monodon* and *P. indicus*. Success has been achieved in obtaining viable larvae of *P. indicus* frozen to -40C and of *P. monodon* frozen to -30C. The frozen *P. monodon* larvae developed further without any abnormalities. Also, safe levels of cryoprotectants have been worked out for the embryos of *L. rohita* and common carp. Further research to develop, vitrification and cryopreservation of carp embryos is in progress (Ponniah, 1997).

8.2 Gene banking

The development of effective protocols for cryopreservation of sperm and embryos will go a long way in support of sustainability in aquaculture. It will also facilitate the establishment of a national facility for gene banking of fish gametes, embryos and DNA material. The existence of such gene bank facilities will be also highly helpful for the obvious reasons that it reduces the maintenance problems and costs involved in raising brood fish in *in situ* condition, provides a wider choice of genetic material for selective breeding, and enables the process of dissemination of genetically improved varieties of fish with ease. Conservation of germplasm of endangered and wild populations would become easier. As mentioned earlier the mini gene bank at NBFGR can be upgraded into a National Fish Germplasm Bank for wild and endangered species. Similar gene banking facilities can be also developed at CIFA for preservation of genetically improved carp and other domesticated culture species.

9. SUMMARY

The Indian major carps, the catla, the rohu and the mrigal are the natural inhabitants of the perennial rivers network of India, Pakistan and Bangladesh and, enjoy a wide distribution. These species are well established in almost all the rivers and reservoirs where they have been transplanted. Substantial research has been carried out to study the biological aspects and propagation of these carps after the initiation of systematic research on inland fisheries in the beginning of 1950s. Based on this information culture technologies were developed resulting in intensive and extensive composite (multispecies) culture; production levels increased from 0.6 t/ha/year to 10-15 t/ha/year. At the same time three decades of effort to develop improved varieties of carps through cross breeding (hybridization) did not result in the production of any commercially viable hybrid varieties.

The Cytogenetic investigations mainly on karyotypes have shown that though these major carps have their own distinct morphological features, genetically they appear to be very closely related as indicated by not only their karyotypes but also by their isozyme and DNA profiles.

The information available on biochemical genetic studies is not very complete. However, recent efforts in this area and in DNA analyses have resulted in the development of some methodologies for the identification of genetic markers in the case of catla, rohu and mrigal. These techniques may be effectively used for stock identification and to study genetic variation in the individuals of a particular species or putative land races. These methods were hitherto not developed and consequently no early studies were conducted to investigate whether there exist different populations or races within a given species of major carps.

With the development of allozyme, RAPD, mtDNA RFLP and microsatellite techniques, some preliminary investigations were carried out which indicated the existence of genetic variations between the stock of different major seed producing hatcheries and also between the hatchery and wild populations of catla.

Differences in isozyme markers have been also noticed among rohu of different river systems namely, Rivers Ganga, Gomati, Yamuna, Sutlej and Brahmaputra.

Of the three species of Indian major Carps, rohu grows relatively slower in carp culture systems, while catla and mrigal exhibit comparatively better growth. Because rohu is a highly prized fish, efforts are being made to improve genetically the species through selection. There appears to be regional preference, for a particular species in India. While rohu is preferred in most of the states in the Eastern region, catla is popular in the Southern states, especially in the provinces of Karnataka and Andhra Pradesh. Genetic improvement of catla has been taken up in Karnataka province.

The aim of current selection projects is to provide correct breeding procedures to the hatchery managers and to develop genetically superior lines that can produce quality seed which can grow to over a kilogram within one year under intensive composite farming system.

The slowest growing among the major carps, *L. calbasu*, is an up-coming species and gaining popularity. As mentioned earlier, this species has good resistance against parasite infection and can stand rough handling to a better extent than other major carp species.

In almost every carp farming nation in Asia, proper use of the genetic potentials of these species has not been sincerely attempted, probably due to the preoccupation to develop suitable culture technologies with the already existing resources and the need to meet the immediate and urgent protein requirement of the masses in their respective nations. Now many of these nations in Asia, after achieving the goal to a notable extent in rising the average fish production levels through improved culture technologies, have started realizing that quality improvement is necessary to increase further the quantity. These two aspects cannot be separable. During the coming decades, if aquaculture is to be a viable industry, it must be supported by technologies which ensure sustainable higher productions. Conservation and protection of the genetic resources and exploitation of genetic potentials of the species of fish to their optimum levels appears to be an excellent immediate path way to achieve the goal.

Most of the nations especially in Asia have already initiated action by taking up conservation of genetic resources and genetic improvement activities of important species. The major carp culturing nations India, Bangladesh, Pakistan, China, Thailand, Vietnam etc. have already started working on the job either independently or in a collaborative manner among themselves or with other advanced nations. The International Network for Genetics in Aquaculture (INGA) is coordinating fish genetic research among all the member countries. The above mentioned countries are among the member countries under the INGA network. India, Philippines and Vietnam have collaboration with the Institute of Aquaculture Research (AKVAFORSK), Norway on selective breeding programmes. Recently the International Centre for Living Aquatic Resource Management (ICLARM) has launched a carp genetic improvement project in which the research on genetic improvement of carps is supported through providing funds and also coordinating the research among the INGA member countries of Bangladesh, China, India, Indonesia, Thailand and Vietnam. Exchange of carp germplasm among interested countries has been also planned under this programme. The coming decade may witness an overall improvement in the aquaculture sector through the systematic and sustainable exploitation of fish genetic resources and the wider use of genetically improved varieties in aquaculture system.

10. REFERENCES

Alikunhi, K.H. 1948. On the occurrence of the mrigal, *Cirrhinus mrigala* Ham. in the Godavari. *Curr. Sci.*, 18(1) : 11-12

Alikunhi, K.H. 1957. Fish culture in India, *Fm Bull. Indian Coun. agri. Res.* (20) : 144

Alikunhi K.H. and H. Chaudhuri. 1951. On the occurrence of *Labeo rohita* in the Godavari river system. *Sci. Cult.*, 16(11): 527.

Alikunhi, K.H. and H. Chaudhuri. 1959. Preliminary observations on hybridization of the common carp (*Cyprinus carpio*) with Indian carps. *Proc. Indian. Sci. Congr.*, (3) : 46.

Alikunhi, K.H., K.K. Sukumaran, and S. Parameswaran. 1963a. Induced spawning of Chinese carps, Ctenopharyngodon idellus (C and V) and silver carp (*Hypophthalmichthys molitrix* in ponds at Cuttack, India. *Proc. IPFC.* 10 (2): 181-204.

Alikunhi, K.H., K.K. Sukumaran and S. Parameswaran. 1971. Studies on composite fish culture. Production by compatible combinations of Indian and Chinese carps. J. Inland Fish. Assoc., 1: 26-57.

Alok, D., N.S. Haque and J.S. Killan. 1995. Transfer of human growth hormone gene into Indian major carp (*L. rohita*) Int. J.Anim. Sci. 10: 5-8.

Andriyasheva, M.A. 1973. Viability of phytophagous fish hybrids. Khozyastva, 85 : 73-86.

Anon. 1974. Fourth Workshop Report of All India Co-ordinated Research Project on Ecology and Fisheries of Freshwater Reservoirs, Ootacamund (Tamil Nadu), 2-3 May, 1974. I.C.A.R. (Mimeo).

Banerjee, S.K. 1987. Induction of fish mitotic division in *vivo* by Concanavalin A. *Indian J. Exp. Biol*, 25: 556-558.

Barat, A., G. John and W.S. Lakra, 1990. Preliminary investigations on G-banding and NOR in some Indian teleosts. Proc. Seventh All India congress of cytology and genetics, December 22-26, 1990, Kalyani, West Bengal Abst. 67.

Barat, (Sahoo), P.K. and A.R. Khuda-Bukhsh. 1992. Electrophoretic patterns of transferrin in some thirty four species of teleostean fishes from India. *In* G.K. Manna and S.C. Roy, eds. (vol.7). *Perspectives in Cytology and Genetics*: 199-211.

Basavaraju, Y. and T.J. Varghese. 1980. Organoleptic evaluation of flesh of two major carp hybrids. *Curr. Res.* 9: 47-48.

Basavaraju, Y. and T.J. Varghese. 1980a. The taxonomic characters of rohu-mrigal and mrigal-rohu hybrids. *Mysore J. Agri. Sci.* 14 (4) : 567-573.

Basavaraju, Y . and T.J. Varghese. 1980b. A comparative study of the growth rate of rohu-mrigal and mrigal-rohu hybrids and their parental species. *Mysore. J. Agric.* Sci., 14 (3): 388-395.

Basavaraju, Y. and T.J. Varghese. 1981. Embryonic and larval development of rohu-mrigal hybrid. *Proc. Indian Acad. Sci.,* 90 (4): 417-426.

Basavaraju, Y. and T.J. Varghese. 1983. Life history of mrigal-rohu hybrid. *Mysore J. agric. Sci.,* 17: 57-65.

Basavaraju, Y., G.Y. Keshavappa, D. Seenappa and K.V. Devaraj, 1989. First report on an interspecific hybrid between *Labeo rohita* and *Labeo fimbriatus*. *Curr. Res.*, 18: 164-165.

Basavaraju, Y., G.Y. Keshavappa, D. Seenappa, K.V. Devaraj and K.V. Mohire. 1990. Growth of *Labeo rohita* x *Labeo fimbriatus* hybrid in comparison with the parental species. The second Indian Fisheries Forum Proceedings, May 27-31, 1990. Mangalore, India. 20-22.

Basavaraju, Y., K.V. Devaraj, S.P. Ayyar and G.Y. Keshavappa. 1994. Morphometric and meristic characteristics of catla (male) x Fimbriatus (female) hybrids. Proceedings of the National Symposium on Aquaculture for 2000 AD. (Ed) Dr. Samuel Paul Raj. 245-252.

Basavaraju, Y. K.V.Devaraj and S.P. Ayyar. 1995. Comparative growth of reciprocal carp hybrids between *Catla catla* and *Labeo fimbriatus. Aquaculture* 129: 187-191.

Basavaraju, Y., G.C. Mair and D.J. Penman, 1997. A genetic improvement programme on *Catla catla* in Karnataka State, India-Preliminary findings. Paper presented at workshop on "Genetic improvement of carps in Asia" held at CIFA, Kausalyaganga during 26-29, July, 1997 by ICLARM/ADB. Philippines.

Basu, S.K., S. Ganguly, S.K. Sarkar and T.K. Basu. 1992. Ontogenetic and Comparative analysis of LDH isozymes in the three species of Indian major carps. *Indian J. Exp. Biol*. 30(2) : 90-93.

Bentsen, H.B., T. Gjedrem, T.M. Thien and N.C. Dan. 1996. Breeding plan for mrigal (*Cirrhinus mrigala*) in Vietnam. Individual (Mass) selection to improve growth rate. Report No. 2. International Network on Genetics in Aquaculture, (INGA).

Bhowmick, R.M., G.V. Kowtal, S.D. Gupta and R.K.Jana. 1981. Some observations on the catla (female) x calbasu (male) hybrid produced by hypophysation. *In* K. Tiews eds. Proceedings, world symposium on Selection, Hybridization and Genetic Engineering in Aquaculture II. Bordeaux, 27-30 May, 1986. Berlin. 102-107.

Bhowmick, R.M., R.K. Jana, S.D. Gupta, G.V. Kowtal and M. Rout. 1981. Studies on some aspects of biology and morphometry of the intergeneric hybrid, *Catla catla* Hamilton x *Labeo rohita* Hamilton. Produced by hypophysation. *Aquaculture*, 23: 367-371.

Burlakov, A.B., A.P. Maneeva, and I.N. Rybarov. 1973. Isozyme composition of hybrids and gynogenetic forms of some cyprinids at early stage of ontogenesis. In Biochemical genetics of fish. 85-90 eds. V.S. Kirpicnikov, A.S. Troshin and I.I. Fridhyanskaya. Institute of cytology. Academic of Sciences USSR.

Chacko, P.I. 1946. A note on carp spawning in the Godavari and Krishna (Madras). *Proc. Indian Sci. Congr.* 33(3): 128.

Chacko, P.I. 1948. Development of Fisheries of the Periyar Lake. *Proc. Indian Sci. Cong.*, 35(3) : 204.

Chacko P.I. and G.K. Kuriyan.1948a. On the bionomics of *Catla catla* (Cuv. and Val.) in South Indian Waters. *Curr. Sci.*, 17: 191.

Chakraborty, R.D. and D.S. Murty. 1972. Life history of Indian major carps, *Cirrhinus mrigala* (Hamilton), *Catla catla* (Hamilton) and *Labeo rohita* (Hamilton). J. Inland Fish. Soc. Ind. 4: 132-61.

Chaudhuri, A. and Gopala Krishna. 1998. Tissue specificity and degree of polymorphism of five enzyme systems of *L. rohita*, from river Yamuna. Fish Gen. Biodiversity Conserv. Natcon Pub. 05: 358-360.

Chaudhuri, H. 1959. Experiments on hybridization of Indian carps. Proc. *Indian Sci. Congr.*, 46 (4) : 20-21.

Chaudhuri, H. 1960. Experiments on induced spawning of Indian carps with pituitary injections. *Indian J.Fish* 7 : 20-48.

Chaudhuri, H. 1967. Breeding and selection of cultivated warm-water fishes in Asia and the Far East. *FAO Fish. Rep.* (44). Vol. 4: 30-66.

Chaudhuri, H., 1971. Fish hybridization in Asia with special reference to India. Lectures, Seminar/Study tour in the USSR, on "Genetic Selection and Hybridization of cultivated fishes". 19 April-29 May 1968. *FAO Rep.* (TA) 2926, 151-159.

Chaudhuri, H., 1973. Fertility of hybrids of Indian carps and preliminary studies on the F2 generations of carp hybrids. *J. Inland Fish. Soc. India*, 5: 195- 200.

Chaudhuri, H. and K.H. Alikunhi. 1957. Observations on the spawning in Indian major carps by hormone injection, *Curr. Sci.* 26, 381-382.

Chaudhuri, H., R.D. Chakraborty, N.G.S. Rao, K. Janakiram, D.K. Chatterjee and S. Jena. 1974. Record production with intensive culture of Indian and Exotic carps. *Curr. Sci.* 43 (10): 303-304.

Chaudhuri, H., R.D. Chakraborty, P.R.Sen, N.G.S. Rao and S. Jena. 1975. A new high in fish production in composite fish culture in freshwater ponds. *Aquaculture*, 6: 343-355.

Chaudhuri, H., N.G.S. Rao, G.N. Saha, M.Rout and D.R. Kanaujia, 1978. Record fish production through intensive fish culture in a farmer's ponds. *J. Inland . Fish. Soc. India*. 10: 19-27.

Cherfas, N.B., Baris Gomelsky, Jaomi Ben-Dom, Yakov Peretz and Gideon Hulata. 1994. Assessment of triploid common carp (*Cyprinus carpio L.*) for culture. *Aquaculture* 127: 11-18.

Chevassus, B. 1983. Hybridization in fish Aquaculture, 33: 245-262.

Child, A.R. and D.J. Solomon. 1997. Observations on morphological and biochemical features of some cyprinid hybrids. *J. Fish Biol.*, 11: 125-131.

Chondar, S.L. 1977. Systematic account of carp hybrid *Labeo gonius* x *Labeo calbasu*. *J. Inland Fish Soc. India*. (9): 172-174.

Chourrout, D. 1980. Thermal induction of diploid gynogenesis and triploidy in the eggs of the rainbow trout (*Salmo gairdineri* Richardson *Nutr. Dev.*, 20 (3A): 727-733.

CIFA/AKVAFORSK. 1996. Final report on selective breeding of rohu (*Labeo rohita*) An Indo-Norwegian collaborative project, May 1992 to March 1996.

Das, P. and A. Barat. 1990. Fish habitat degradation necessitating conservation. *Environmental Series* Vol. *IV*: 85-89.

Das, P. and A. Barat. 1995. *In* G.K. Manna & S.C. Ray eds. Application of genetics in fisheries can help blue revolution in India. Perspectives in Cytology and Genetics. 8 (1995) : 25-33.

David, A., P. Ray, B.V. Govind, K.V. Rajagopal and R.K. Banerjee. 1969. Limnology and fisheries of the Tungabhadra reservoir *Bull. Cent. Inl. Fish. Res. Inst.*, Barrackpore. (13) : 188 (Mimeo).

Day. F. 1878. The fishes of India being a natural history of the fishes known to inhabit the seas and freshwaters of India, Burma and Ceylon. London Wilham Dawson and Sons, 778. (Reprinted 1958).

Day F. 1889. The fauna of British India, including Cylon and Burma. *Fishes 1, London*, pp 287.

Desai, V.R. and K.J. Rao. 1970. On the occurrence of natural hybrid of catla-rohu in Madhya Pradesh. *J.Zool. Soc. India*, 22 (1 & 2) : 35-40.

De Witt, H.H. 1960. A contribution to the ichthyology of Nepal. *Stanford Ichthyal. Bull.*, 7(4) : 63-88.

Dubey, G.P. 1969. Induced breeding of Indian carp in dry bundhs. In FAO/UNDP Regional Seminar on induced breeding of fishes, Calcutta, Rome, FAO, FRC/IBCF/21: 10.

Dubey G.P. and R.P. Tulu. 1961. Observations on the breeding of major carps in Madhya Pradesh. *J.Bombay Nat. Hist. Soc.*, 58(1): 81-91.

Eknath, A.E. and R.W. Doyle. 1985. Indirect selection for growth and life history traits in Indian carps aquaculture. 1. Effects of brood stock management +*Aquaculture*, 49: 73-84

Eknath, A.E. and R.W.Doyle. 1990. Effective population size and rate of inbreeding in aquaculture of Indian major carps. *Aquaculture*, 85 : 293-305.

Eknath, A.E. 1991. Simple broodstock management to control indirect selection and inbreeding. Indian carp example. *Naga*, April, 1991. The ICLARM contribution No. 738 : 13-14.

FAO, 1972. Report of the first meeting of the FAO *Ad hoc* working parts of Genetic Selection and the conservation of Genetic resource of fish. Rome, Italy, 7-10 Dec., 1971. FAO Fish Rep, (119) : 9.

FAO Database on Introductions of Aquatic Species (DIAS)..

FAO, 1980. Conservation of the genetic resources of fish. Problems and recommendations. Report of the Expert Consultation on the Genetic Resources of Fish. Rome, Italy 9-13 June : 43.

Frankel, O.H. and M.E. Soule. 1981. Conservation and evolution. Cambridge University Press, Newyork, N.Y.

Ganapati, S.V. and K.H. Alikunhi. 1950. Notes on the spawning of carps in River Tungabhadra in response to off season freshets. *J.Zool. Soc. India*. 2: 93-5.

Gjarde, B., P.V.G.K.Reddy, K.D. Mahapatra, J.N.Saha, R.K.Jana, S.D. Gupta, M. Sahoo, S. Lenka and M. Rye. 1998. Complete diallel cross between rohu stocks (*L. rohita*). In MSS.

Gjarde, B., P.V.G.K.Reddy, S.D.Tripathi, R.K.Jana, K.D. Mahapatra, S.D. Gupta, J.N.Saha, S. Lenka, M. Sahoo, P. Govindassamy, M. Rye and T. Gjedrem. 1998. Selective breeding of rohu in India. (In Press).

Gold, J.R. 1979. Cytogenetics. *In* W.S. Hora, D.J.Randal and J.R. Brott, eds. In Fish Physiology , Vol. 8 . 353-405. New York, Academic Press.

Gold, J.R., Y.C.Li, N.S. Shipley and P.K. Powers. 1990. Improved methods for working with fish chromomomes with a review of metaphase chromosome banding. *J.Fish. Biol*, 37: 563-575.

Goldschmidt, C.R. 1953. Catla fishing in powai Lake, Greater Bombay. *J.Bombay Nat. Hirt. Soc.*, 55(3) : 473-480.

Golovinskaia, K.A. 1968. Genetics and selection of fish and artificial gynogenesis of the carp (*Cyprinus carpio*) FAO Fish. (44), Vol. 4: 215-222.

Gopal, C.K. T.J.Varghese and P. Keshavanath. 1989. *In* . Das and Jhingran eds. Morphological features of catla-mrigal and Mrigal- catla hybrids and their comparison with the parental species. Fish Genetics in India, (1989): 183-191. Today and Tomorrow's Printers and Publishers, New Delhi-110 005. (India).

Gui, J.Y: Li, K. Li, and T. Zhou, 1986. Trans. Chin. Ichthyology Soc. V. Sci. Press. Beijing. 115-123.

Gupta, M.V. and B.O. Acosta. 1996. Proceedings of the third INGA steering committee meeting. Cairo, Egypt. 8-11 July, 1996.

Gupta, S.D. 1994. Procurement of rohu stocks from different river systems of India and their rearing and management to raise brood fish. In the regional workshop manual on "Selective breeding of rohu, *L. rohita*" (an Indo-Norwegian collaborative Project) June 28 - July 12, 1994.

Gupta, S.D. and S.C. Rath. 1993. Cryogenic Preservation of carp milt and its utilization in seed production. The third Indian Fisheries Forum Proceedings 11-14 October, 1993, Panthnagar. 77-79.

Gupta, S.D. S.C. Rath and S. Dasgupta. 1995. Fertilization efficiency of cryopreserved carp spermatozoa over four years at different time intervals after thawing. *Geobios* 22 : 208-211.

Gupta, S.D., S.C. Rath, S. Dasgupta and S.D. Tripathi. 1995. A first report on quadruple Spawning of *Catla catla (Ham.) Vet. arhiv* 65(5), 143-148.

Guyomard, R. 1978. Identification par electrophorese d ' hybrides de salmonides Ann. Gen. Sel. Anim., 10: 17-27 (in French with an English Summary)

Hora, S.L. 1945. Symposium on the factors influencing the spawning of Indian carp. *Proc. natri. Inst. Sci. India* (B) 11: 303-12.

Hornel, J. 1924. Administrative report for the year 1922-27. *Madras Fish. Bull.*, 18: 1-58.

Hubbs, C.L. 1955. Hybridization between fish species in nature. *Syst. Zool.*, 4(1): 1-20.

Husain, A. 1945. Analysis of experimental data on the spawning of carp in the Punjab. *Proc. nat. Inst. Sci. India* (B), 11: 320-4.

Hussain, M.G., M.S, Islam., M.S. Rahman and S.C. Mahata. 1994b. Suppression of first cell division and production of mitotic gynogens in rohu. *Laboe rohita* Ham. Poster paper presented at the fifth International Symposium on Genetics in Aqua. 19-25 June 1994. Dalhousie University, Halifax. 22pp.

Hussain, M.G., G.P.S. Rao, N.M. Humayun, C.F. Randall, D.J. Penman, D. Kime, N.R. Bromage, J.M. Myers and B.J. Mc Andrew. 1995. Comparative performance of growth, biochemical composition and endocrine profile in diploid and triploid tilapia. Oreochromis miloticus L., *Aquaculture* 138: 87- 97.

Hussain, M.G. and M.A. Mazid. 1997. *In* M.V. Gupta, M.M. Dey, R. Dunham and G. Bimbao Carp genetic resources in Bangladesh. In proceeding of the collaborative research and training or genetic improvement of carp species in Asia, 26-29 July 1997. Central Institute of Freshwater Aquaculture, Bhubaneswar, India.

Hussain, M.G., S.C. Mahata, M.S. Rahaman, M.B. Tanu, M.A. Mazid and M.S. Islam. 1997. Induction of mitotic and Meiotic gynogenesis and production of genetic clones in rohu, *L. rohita* Ham. *Bangladesh J. Fish Res.*, 1(2): 01-07.

Ibrahim, 1977. Contribution to hybridization among Indian carps. Studies on the biology and morphological features of some Indian carp hybrids. Thesis for the degree of Doctor of Philosophy of the Utkal University.

Ibrahim, K.H., G.V. Kowtal and S.D. Gupta. 1980. Embryonic and larval development among *Catla catla* (Hamilton) x Hypophthalmichthys molitrix (Valenciennes) hybrid. *J. Inland Fish Soc. India.* Vol. 12. No. 2: 69-73.

Ibrahim, K.H., S.D. Gupta and P.R. Sen. 1982 . Malformation of the vertibral column in single spawn progeny of *L. rohita* (Hamilton). Bamidgeh. 34 (2): 59-62.

Islam, M.S., M.S. Shah, M.A. Rahman and H.A. Chaudhury. 1994. Induction of triploidy in rohu, *L. rohita* by hear shock treatments and comparative growth with normal diploids. J. Aqua. Trop, 9: 299-310.

Jana, R.K. 1993. Studies on some aspects of biology and cytogenetics of rohu-catla and catla-rohu hybrids. Thesis submitted for *the degree of Doctor of philosophy (Science) of the University of Calcutta.*

Jhingran, A.G. 1984. The Fish Genetic Resources of India. Indian Council of Agricultural Research, New Delhi.

Jhingran, A.G. and R.A. Gupta. 1989. *In* Das & Jhingran, eds. The relevance of selective breeding for genetic improvement of formed fish. *Fish genetics in India*. 135-140. Today and Tomorrow"s Printers and Publishers, New Delhi (India).

Jhingran. V.G. 1968. Synopsis of Biological data on catla, *Catla catla* (Ham. 1822). FAO Fisheries Synopsis No. 32 Rev.1.

Jhingran, V.G. and A.V. Natarajan. 1969. An assessment of the D.V.C. reservoirs in relation to stocking. In seminar on the *Ecology and Fisheries of Freshwater Reservoirs*. Barrackpore, November 27-29, 1969.

Jhingran V.G. and H.A. Khan. 1979. Synopsis of Biological data on the mrigal, *Cirrhinus mrigala* (Hamilton 1822). FAO Fisheries Synopsis No. 120. 78p.

Jhingran, V.G. 1982. Fish & Fisheries of India. Revised edition . Hindustan Publishing Corporation (India), G-U-B, Jawahar Nagar, New Delhi. 666.

Job T.J. and P.I. Chacko. 1948. On the breeding grounds of catla and other major carps in the South Indian rivers. *Proc. Indian Sci. Congr*. 34(7) : 183.

John, G., P.V.G.K.Reddy and S.D. Gupta. 1984. Artificial gynogenesis in two Indian major carps, *Labeo rohita* (Ham.) and *Catla catla* (Ham.), *Aquaculture*, 42: 161-168.

John, G. and P.V.G.K. Reddy. 1986. A note on the *Labeo rohita* (Ham.) x *Cyprinus carpio* hybrid. Proc. World Symp. on selection, Hybridization and Genetic Engineering in Aquaculture. Bordeaux 27-30 May 1986. Vol 11. Berlin.

John, G., P.V.G.K.Reddy and R.K. Jana. 1988. Induced gynogenesis in the Indian major carp, *Cirrhinus mrigala* (Ham.). In Mohan Joseph. (Ed.), the first Indian Fisheries Forum. Proceedings. Asian Fisheries Society. Indian Branch, Mangalore. 107-109.

John, G., A. Barat and W.S. Lakra. 1992. Localisation of nucleolar organizer regions in fish species. *Catla catla* and *Mystus vittatus. The Nucleus* 35: 179- 181.

John, G., A. Barat and W.S. Lakra. 1993. Localisation of nucleolar organizer regions in *Labeo* (Cyrpinidae). *La Kromosomo* II-70 : 2381-2384.

John, G., A.G. Ponniah, W.S. Lakra, A. Gopalakrishnan and A. Barat. 1993. Preliminary attempts to cryopreserve embryos of *Cyprinus carpio* (L/ and *L. rohita* (Hamilton). J. Inland Fish Soc. India, 25 (1): 55-57.

Keshavanath, P., T . J. Varghese and P. K. Reddy. 1980. Preliminary studies on the growth of catla-rohu and rohu-catla hybrids. *Mysore J. Agri. Sci*. 14: 401-407.

Khan H. 1947. Observations on the spawning of carp in a tank in Punjab. Proc. *Indian Sci. Congr*. 34(3) : 184.

Khan H.A. and V.G. Jhingran. 1975. Synopsis of Biological data on Rohu *Labeo rohita* (Ham., 1822). FAO Fisheries Synopsis No III. 100.

Khan, H.A., S.D. Gupta and P.V.G.K.Reddy. 1988. Suitable dye for marking Asiatic carps. *Indian Journal of Animal Sciences* 58 (8): 1003-1006.

Khan, H.A., P.V.G.K.Reddy and S.D.Gupta. 1989. Studies on some aspects of biology and morphology of the intergeneric hybrid. *Cyprinus carpio* var. *Communis L.* x *Labeo rohita* (Hamilton). *Proc. Nat Acad. Sci. India*. 59(B), II. 155-162.

Khan, H.A. and G.V. Kowtal. 1989. *In* Das & Jhingran eds. An account of naturally occurring and artificially produced cyprinid hybrids in India. *Fish Genetics in India*, (1989) : 155-162. Today and Tomorrow's Printers and Publishers, New Delhi-110 005 (India).

Khan, H.A., S.D. Gupta, P.V.G.K.Reddy, M.S.Tantia and G.V. Kowtal. 1990. Production of sterile intergeneric hybrids and their utility in Aquaculture and reservoir stocking. 41-48. *In* P.Keshavanath and K.V. Radhakrishna eds. Carp seed Production Technology. Proceedings of the workshop on carp seed production technology, 2-4 September, 1988. *Special Publication* 2, pp 102. Asian Fisheries Society, Indian Branch, Mangalore, India.

Khanna, D.V. 1958. Observations on the spawning of the major carps at a fish farm in the Punjab. *Indian J. Fish.*, 5(2): 282-90.

Khuda-Bukhsh, A.R. and G.K. Manna. 1974. Somatic chromosomes in seven species of teleosteam fishes. C 15. 17: 5-6.

Khuda-Bukhsh, A.R. and G. K. Manna, 1976. Chromosome analysis of twc Indian hybrid carps. *Proc. 63rd Ind. Sc. Cong*. 7, 151.

Khuda-Bukhsh, A.R., K. Nayak and P.K. Sahoo. 1988. Cytogenetic indications of matroclinous inheritance in F1 individuals of *Cyprinus carpio var. communis* x *Labeo calbasu*. *Proc. Zool. Soc.* Calcutta, 41: 9-17.

Khuda-Bukhsh, A.R. and C.Chakraborty. 1994. Localization of C-band heterochromatin in metaphase chromosomes of two species in Indian major carps. *Inland Fish. Soc. India*, 26 : 44-46.

Khuda-Bukhsh, A.R. and S. Tiwary. 1994. Study of G-bands in two species of fish. *Anabas testudineus* and *Labeo rohita*. Inland Fish Soc. India, 26 : 47-50.

Khuda-Bukhsh, A.R. and P.K.Sahoo. 1998. *In* G.K. Manna and S.C.Roy, AICCG Pubt, Kalyani University eds.Gel electrophoretic studies of plasma protein profiles in thirtyfive odd species of Indian teleostean fish and an assessment of their importance in biosystematics. *Perspectives in Cytology* and *Genetics*. 1998, 9: 139-157.

Konda Reddy, P. 1977. Studies on taxonomic characters and growth of catla-rohu and rohu-catla hybrids. M.F.Sc. Thesis, University of Agricultural Science, Bangalore, 136 .

Konda Reddy, P and T.J. Varghese. 1980. Embryonic and larval development of the hybrid, Catla-rohu. *Mysore J. Agri. Sci.*, 14 : 588-598.

Konda Reddy, P. and T.J. Varghese. 1980a. On the taxonomic characters of two major carp hybrids, catla-rohu and rohu-catla. *Proc. Indian Acd. Sci*. (Anim. Sci.), Vol. 89; (5) : 419-429.

Konda Reddy, P., T.J. Varghese. 1980b. Studies on the comparative growth rate of two Indian major carp hybrids, catla-rohu and rohu-catla. *J. Inland Fish Soc. India*, Vol. 12. No. 2: 18-24.

Konda Reddy, P. and T.J. Varghese. 1983. Life history of a major carp hybrid rohu-catla. *Indian J.Fish*. 30 (1): 87-95.

Kowtal, G.V. and S.D. Gupta. 1985. A note on the hybrid mrigal, *Cirrhinus mrigala* (Ham.) x Common carp, *Cyprinus carpio* L. *Aquaculture*, 49: 179-183.

Krasznai, Z., T. Marian. 1986. Shock induced triploidy and its effect on growth and gonad development of European catfish (*Silurus glanis*) L. J.Fish Biol. 29: 519-527.

Krishnaja, A.P. and M.S. Rege, 1975. Cytogenetic studies on two species of genus Labeo and their fertile hybrid. Proc. 2nd All India Cong. Cytol. & Genet., J. Cytol. & Genet. Suppl. 125.

Kuldeep, K. Lal, P.V.G.K. Reddy, Birbal Singh, J.N. Saha, V. Mohindra, K.D. Mahapatra, R.K.Jana, B.Gjerde, Morten Rye and A.G. Ponniah, 1998. Genetic characterization of *L. rohita* founder stocks used in selective breeding programme. (unpublished).

Kumar, K. 1989. Das & Jhingran eds. Studies on cryogenic preservation of carp spermatozoa. *Fish Genetics in India*, (1989) : 199-206, Today & Tomorrow's Printers and Publishers, New Delhi-110 005 (India).

Mahapatra, K.D., P.V.G.K. Reddy, J.N.Saha, S.D. Gupta, R.K.Jana and Smita Lenka. 1998. *In* P.C. Thomas ed. Prod. Current and Emerging Trends in Aquaculture. 191-194. Editor; P.C. Thomas, Daya Publishing House, 1998, New Delhi-110 035.

Maheshwari, U.K. and R.S. Biradav, 1998. Decline in fertility of hatchery bred *Labeo rohita* due to inbreeding. *Proc. Fish Gen., Biodiversity conserv. Natcon Pub.* 05. 412-416.

Mair, G.C., A.G. Scott., J.A. Beardmore and D.O.F. Skibinski. 1986. A technique for induction of diploid gynogenesis in *Oreochromis niloticus* by suppression of the first mitotic division. Proceedings of EIFAC/FAO Symposium on Selection, hybridization and genetic engineering in aquaculture of fish and shelfish for consumption and stocking. Bordeaux (France) 27-30 May 1986.

Malison, J.A., L.S. Procarioue, J.A. Held, T.B Kayes and C.H. Amundson. 1993. The influence of triploidy and heat and hydrostatic pressure shocks on the growth and reproductive development of Juvenile yellow perch (*Perca flavescens*), *Aquaculture*. 116: 121-133.

Majumdar, K. and S.P. Ray Chaudhuri, 1976. Studies on the chromosomes of Indian major carps. *Proc. Sym.* "Modern treads in Zoological Researches in India" Calcutta, July 26-27. Abs. 68-69.

Majumdar, K.C., K. Ravinder and K. Nasaruddin. 1997. DNA fingerprinting in Indian major carps and tilapia by Bkm 2 (8) and M 13 probes. *Aquaculture Research* 28: 129-138.

Manna, G.K. and R. Prasad, 1971. A new perspective in the mechanism of evolution of chromosomes in fishes. *J. cytol. Genet*. 239-240. Proc. 1st All India Cong. Cytol. & Genti.

Manna, G.K. and A.R. Khuda-Bukhsh. 1977a. Karyomorphology of cypriniod fishes and cytological evoluation of the family. *The Nucleus* 20 (1 & 2): 119-127.

Manna, G.K. and A.R. Khuda-Bukhsh. 1977. A check list of chromosomes in cyprinid fishes. *J. Zool. Res.*, Vol. 1, No. 2. p. 34-43.

Mantelman, II, 1973. Cytological investigations of several phytophagous fishes and their hybrids. *Ozemogo Khozyastva*, 85: 87-93.

Marian, T. and Zoltan Krasznai. 1978. Karyological investigations on *Ctenopharyngodon idella* and *Hypophthalmichthys nobilis* and their cross-breeding. *Aquaculture Hangarica* (Szarvas). Vol. 1: 44-50.

McAndrew, B.J., K.J. Rana and D.J. Penman. 1994. *In* Recent advances in Aquaculture eds. Conservation and preservation of genetic variation in aquatic organisms. James F. Muir and Ronald J. Roberts. IV: pp 297-337.

Menon M.D., S. Srinivasan and B.Krishnamurthi, 1959. Report to the Indian Council of Agriculture Research on the Madras rural piscicultural scheme worked from 1 July 1942 to 31 March 1952. Madras, Govt. Press, 171.

Mishra K.S. 1959. An aid to the identification of the common commercial fishes of India and Pakistan. *Rec. Indian Mus.*

Mishra, B.K. 1991. Observations on the susceptibility of the hybrid, *L. rohita* x *Catla catla* to *Argulus siamensis* (Wilson) infection in composite culture ponds. In proceedings of National symposium on Freshwater Aquaculture. 84-88, Bhubaneswar, India.

Mishra, A. and R.P. Raman. 1993. Genetic issues involved in fish seed production and management in India. *Fishing Chimes*. March, 43-45.

Mishra, A., A.K. Jain and S.P. Singh. 1998. Rate of spontaneous hybridization in "mixed spawning" of Indian major carps. *Fish. Gen. Biodiversity Conserv. Natcon Pub.* 05 : 417-428.

Nagpure, N.S. 1997. Distribution of C-band heterochromatin in *Labeo rohita* Ham. (Cyprinidae). *Chromosome Science* 1: 45-46.

Nagpure, N.S. and A. Barat. 1997. A simplified method of fish chromosome preparation by *in vitro* colchicine treatment. *Indian Journal of Experimental Biology* Vol. 35: 915-916.

Nagy, A., V. Csanyi, J. Bakos and M. Bercsenyi. 1984. Utilization of gynogenesis and sex reversal in commercial carp breeding. Growth of the first gynogenetic hybrids. *Aquaculture Hungarica* (Szarvas) 4: 7-16.

Nasar, S.K.T. and H. Haque. 1989. *In* Das & Jhingran eds. Strategy for genetic preservation. An overview. *Fish Genetics in India.*, 207-216. Today & Tomorrow. Printers and Publishers, New Delhi (India).

Naseem Hamsa. K. 1971. Preliminary account of an intergeneric fertile hybrid between *Cirrhinus mrigala* and *Labeo rohita. Annual Day Souvenir*, CIFE., Bombay, 10-15 Naseem Hamza, K., 1972. Studies on some cyprinid hybrids. Ph.D. Thesis. Bombay University, Bombay.

Naseem Hamsa and K.H. Alikunhi. 1971. Preliminary studies on upgrading *Cirrhinus reba* by hybridization. *J. Indian Fish. Assoc.*, 1 (1): 68-73.

Natarajan, A.V. and A.G. Jhingran. 1963. On the biology of *Catla catla* (Ham.) from the river Yamuna. *Proc. natn. Inst. Sci. India* (B), 29(3) : 326-55.

Natarajan, A.V., V.R. Desai and D.N. Mishra. 1976. On the natural occurrence of the intergeneric catla-rohu hyb[illegible] Rihand (Uttar Pradesh) with an account of its potential role in reservoir fisheries development in India. *J. Inla[illegible] Fish. Soc. India*. Vol. 8 : 83-90.

Nuanmanee Pongthana, 1997. *In* M.V. Gupta, M.M. Dey, R.Dunham and G. Bimbao eds. Carp genetic resources Thailand. Proceedings of the collaborative research and training on genetic improvement of carp species in Asia, 26-29 July 1997. Central Institute of Freshwater Aquaculture, Bhubaneswar.

ODA Carp Genetics Project Research Progress. 1996. *In* Y. Basavaraju, Graham C. Mair and David J. Penman eds. Characterisation and improvement of *Catla catla*. In Proceedings of the workshop on "genetic improvement of carps".

Padhi, B.K. and A.R. Khuda-Bukhsh. 1989a. Malate dehydrogenase isozyme patterns during the development of five species of cypribnid fishes. *Environment and Ecology* 7(2): 415-419.

Padhi, B.K. and A.R. Khuda-Bukhsh. 1989b. Lactate dehydrogenase isozyme pattern in developing stages of four species of Indian carps (Cyprinidae, Pisces). *Nat. Acad. Sci. Letters*, Vol. 12, No. 2: 59-61.

Padhi, B.K. and A.R. Khuda-Bukhsh. 1990a. Xanthine dehydrogenase isozymes in twenty species of teleostian fishes. Tissue distribution and possible taxonomic significance. *Biochemical systematics and Ecology*. 18(5) : 381-386.

Setna, S.B. and C.V. Kulkarni. 1949. The freshwater fish and fisheries of Ahmedabad. *J. Bombay Nat. Hist. Soc.,* 46(1): 126-32.

Sinha, V.R.P. and H.A. Khan, 1989. *In* Das & Jhingran eds. Genetic improvement of Indian major carps for aquaculture industry. Fish Genetics in India, (1989): 141-146. Today and Tomorrow's Printers and Publisher, New Delhi-110 005 (India).

Slastentenko, E.P. 1957. A list of natural fish hybrids of the world. *Hydrobiology Istanbul.* 4 : 76-97.

Sola, L., B. Camerini and S. Cataudella. 1984. Cytogenetics of Atlantic eel. C- and G-banding, nucleolus organizer regions and DNA content. Cytogent. Cell. Genet. 38 : 206-210

Somalingam, J., U.K. Maheshwari and R.K. Langer. 1990. *In*. P. Keshavanath and K.V. Radhakrishnan eds. Mass production of intergeneric hybrid catla (*Catla catla* O x *Labeo rohita* O) and its growth in ponds small and large reservoirs of Madhya Pradesh.. 49-52. Carp seed production technology. Proceedings of the workshop on carp seed production technology. 2-4 September, 1988. Special publication 2., 102 . *Asian Fisheries Society, Indian Branch. Mangalore, India.*

Sreenivasan, A. 1996. Limnology of tropical impoundments, Hydrobiological and fish production in Stanley reservoir, Mettur Dam. *Inst. Rev. Ges.* Hydrobiology, 51 (2): 295-306.

Stanley, J.G.. 1976. Production of hybrid androgenetic and gynogenetic grass carp and carp. *Trans. Am. fish. Soc.* 105: 10-16.

Stanley, J.G., J.M. Martin and J.B. Jones. 1975. Attempts to produce monosex fishes by artificial gynogenesis. *Progre. Fish Cult.* 37, 25-26.

Stanley, J.G., C.J. Biggers and D.E. Schultz.. 1976. Isozymes in androgenetic and gynogenetic white amur, gynogenetic carp and carp-amur hybrids. *J. Hered.* 67: 129-134.

Sugunan, V.V. 1995. Reservoir fisheries of India. FAO Fisheries Technical paper 345: 423p.

Supsuksamran, M. 1987. Preliminary study on genetics of *L. rohita* (Biochemistry). M.Sc. Thesis, Kasetsart University. 90.

Swaminathan, M.S. 1984. *In Genetics, New Frontiers*, Vol.1. eds. Chopra, V.L., Joshi, B.C., Sharma, R.P. and Bansal, H.C., Oxford and IBH Publishing Co. Ltd. New Delhi, Bombay, Calcutta.

Talwar, P.K. and A.G. Jhingran. 1991. Inland Fisheries of India and adjacent countries. Vol 1 : 541 p. Oxford & IBH publishing co. PVT. LTD. New Delhi, Bombay, Calcutta.

Tripathi, S.D. and R.K. Sharof, 1976. A natural rohu x mrigal hybrid from Adhartal lake, Jabalpur. *J. Zool. Soc.* India, 28, (1 & 2).

Tripathi, S.D., N.K. Sinha and R.K. Sharof, 1974. Hybrid carp from Adartal Lake, Jabalpur. *JNKVV Res. J.*, 8 : (3-4) 165-169.

Tripathi, S.D., P.K. Aravindakshan, S. Ayyappan, J.K. Jana, H.K. Muduli, Suresh Chandra and K.C. Pani. 1997. New Dimensions in intensive carp culture in India. Communicated to Aquaculture, Elsevier Science Publishers B.V., Amsterdam- The Netherlands.

Varghese, T.J. and K.K. Sukumaran. 1971. Notes on hypophysation of catla-rohu hybrid. *Annual Day Souvenir*, CIFE, Bombay. 16-17.

Varghese, T.J. and B. Shantharam. 1979. Preliminary studies on the relative growth rate of three Indian major carp hybrids. *Proc Indian Acad*. Sci. B (3): 209- 216.

Varghese, T.J., P. Keshavanath, M.H. Neghur, P. Konda Reddy and B.G. Mahadevappa, 1984. Evaluation of two major carp hybrids, catla-rohu and catla-mrigal through composite fish culture. *Indian J. Anim. Sci*. 54: 1158-1162.

Wu, C., Y. Ye and R. Chen. 1986. Genome manipulation in carp (*Cyprinus carpio L.*) *Aquaculture*, 54: 57- 61.

Zhang, S.M. 1990. Inducing polyploidy and Karyological studies in Indian major carps *Catla catla* (Ham.), *Labeo rohita* (Ham.) and *Cirrhinus mrigala* (Ham.) and in eight other freshwater species of fish. A Report. The secondment of young Scientist Programme. (April 19, 1989-April 18, 1990). CIFA (ICAR).

Zhang, S.M. and P.V.G.K. Reddy. 1991. On the comparative karyomorphology of three Indian major carps, *Catla catla* (Ham.), *Labeo rohita* (Ham.) and *Cirrhinus mrigala* (Ham.) *Aquaculture*. 97: 7-12.